WORKBOOK

PEARSON EDEXCEL A-LEVEL

Geography

PHYSICAL GEOGRAPHY

Topics 1–2B and 5–6

- Tectonic processes and hazards
- Glaciated landscapes and change
- Coastal landscapes and change
- The water cycle and water insecurity
- The carbon cycle and energy security

HODDER
EDUCATION
AN HACHETTE UK COMPANY

The Publishers would like to thank the following for permission to reproduce copyright material.

Photo credits

p. 37 duncanandison/Adobe Stock, p.59 © David Holmes.

Every effort has been made to trace all copyright holders, but if any have been inadvertently overlooked, the Publishers will be pleased to make the necessary arrangements at the first opportunity.

Although every effort has been made to ensure that website addresses are correct at time of going to press, Hodder Education cannot be held responsible for the content of any website mentioned in this book. It is sometimes possible to find a relocated web page by typing in the address of the home page for a website in the URL window of your browser.

Hachette UK's policy is to use papers that are natural, renewable and recyclable products and made from wood grown in well-managed forests and other controlled sources. The logging and manufacturing processes are expected to conform to the environmental regulations of the country of origin.

Orders: please contact Hachette UK Distribution, Hely Hutchinson Centre, Milton Road, Didcot, Oxfordshire, OX11 7HH. Telephone: +44 (0)1235 827827. Email education@hachette.co.uk. Lines are open from 9 a.m. to 5 p.m., Monday to Friday. You can also order through our website: www.hoddereducation.co.uk

ISBN: 978 1 3983 3243 0

© David Holmes and Michael Witherick 2021

First published in 2021 by

Hodder Education,
An Hachette UK Company
Carmelite House
50 Victoria Embankment
London EC4Y 0DZ

www.hoddereducation.co.uk

Impression number   10 9 8 7 6 5 4 3 2

Year   2025 2024 2023

Cover photo © anekoho – stock.adobe.com

Illustrations by Integra Software Services Pvt. Ltd., Pondicherry, India.

Typeset by Integra Software Services Pvt. Ltd., Pondicherry, India.

Printed by CPI Group (UK) Ltd, Croydon CR0 4YY

A catalogue record for this title is available from the British Library.

Contents

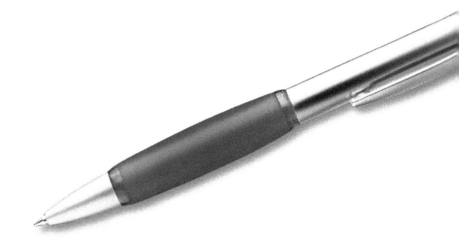

About this book

1   This workbook will help you to prepare for paper 1 of the Pearson Edexcel A-level Geography exam, but will also include links to the skills needed for paper 3.

2   The paper 1 exam is 2 hours and 15 minutes long and includes a range of questions. The exam is divided into three sections:

- Section A: Tectonics is compulsory (16 marks)
- Section B: Landscape Systems involves a choice of either glaciation or coasts (40 marks)
- Section C: Physical Systems is compulsory (49 marks)

3   Each section includes a variety of question types:

- short, point-marked answers
- levels-marked responses
- extended prose

4   The questions in this workbook are scaffolded — they begin with easier, more accessible questions and then work up to more complex questions. Each section ends with questions that are exam-style, some of which encourage you to think about links across different topics, bringing together all your skills and knowledge.

5   You still need to read your textbook and refer to your revision guides and lesson notes.

6   Marks and assessment objectives are indicated for exam-style questions so that you can gauge both the detail and balance of content required in your answers.

The assessment objectives (AOs) are as follows:

- AO1: Demonstrate knowledge and understanding of places, environments, concepts, processes, interactions and change, at a variety of scales.
- AO2: Apply knowledge and understanding in different contexts to interpret, analyse and evaluate geographical information and issues.
- AO3: Use a variety of relevant quantitative, qualitative and fieldwork skills in order to investigate questions and issues, interpret, analyse and evaluate data and evidence, and construct arguments and draw conclusions.

7   Worked answers are included throughout the practice questions to help you understand how to gain the most marks.

8   Icons next to the question will help you to identify:

 where questions draw on synoptic knowledge, i.e. content from more than one topic

 where geographical skills are tested

 where your calculations skills are tested

 how long this question should take you

9   Timings are given for the exam-style questions to make your practice as realistic as possible.

10   Many of the exam-style questions will require more space than is available in this workbook. Answer these on a separate sheet of paper.

11   Answers are available at: www.hoddereducation.co.uk/workbookanswers.

Topic 1 Tectonic processes and hazards

Why are some locations more at risk from tectonic hazards?

Hazards are natural events that have an adverse impact on people, the economy and society. Tectonic hazards include earthquakes and volcanic eruptions, as well as the secondary impacts resulting from the primary event.

The theory of plate tectonics is based on evidence that the Earth's crust is made up of slabs, moved by convection currents in the underlying mantle; it attempts to explain plate movements. There are other theoretical frameworks that attempt to explain plate movements, relating not only to convection currents but also to the role of gravity (leading to slab-pull).

Tectonic hazards have a particular distribution linked to plate boundaries and physical processes operating at these locations. For example, approximately 70% of all earthquakes are found around the Pacific Ring of Fire, surrounding the Pacific Ocean.

Physical processes explain the causes of tectonic hazards. In the case of earthquakes, these are often linked to energy release and the build-up of tectonic strain, followed by a sudden release that causes ground-shaking. Volcanoes present different risks, including pyroclastic flows, tephra and volcanic gases.

Practice questions ?

1   What is meant by the term plate boundary?

....

....

2   Describe the difference between divergent and convergent plate boundaries.

....

....

....

3   Name the **four** types of plate boundary.

....

....

....

4   Identify the **two** plate boundaries where the most powerful earthquakes occur.

....

....

5   Study Figure 1.1.

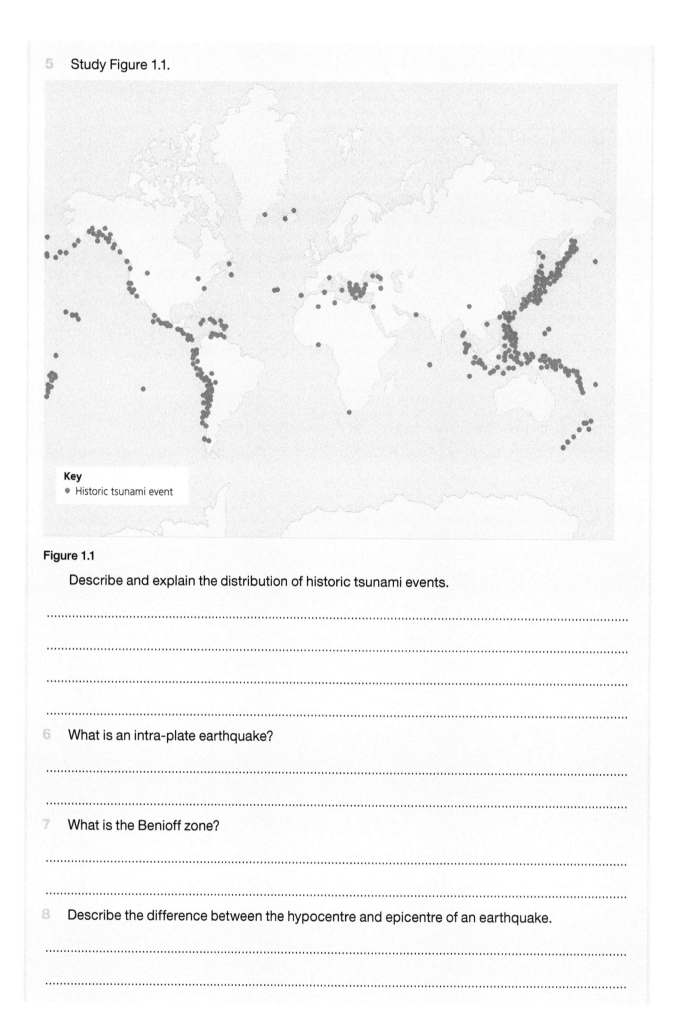

Key
● Historic tsunami event

Figure 1.1

Describe and explain the distribution of historic tsunami events.

....

....

....

....

6   What is an intra-plate earthquake?

....

....

7   What is the Benioff zone?

....

....

8   Describe the difference between the hypocentre and epicentre of an earthquake.

....

....

9　What are hot spot volcanoes and where do they occur?

....

....

....

10　Explain the difference between seismic and volcanic in terms of tectonic classification.

....

....

....

11　Why are some locations more at risk from tectonic hazards than others?

....

....

....

....

12　Describe the differences between the **two** different types of the Earth's crust.

....

....

....

....

13　Explain the different processes that cause tectonic plates to move.

....

....

....

....

14　What is the difference between primary and secondary hazards, and why might their impacts vary?

....

....

....

....

15 Why are there differences in the type and scale of hazards presented by volcanoes?

....

....

....

....

....

16 Explain how different types of seismic wave will likely have different impacts on people and property.

....

....

....

....

....

17 Explain the idea of a locked fault and why it is a cause for concern, especially in relation to energy release. Use some examples in your answer.

....

....

....

....

....

18 Explain why earthquakes and volcanoes have similar, but also different, distributions.

....

....

....

....

....

19 Explain the importance of different factors that influence the impact of a tsunami.

Worked example

The most important physical factor might be the magnitude of the undersea event (eruption/landslide/earthquake) and the resulting displacement of water. Tsunami intensity is measured by run-up height, or the maximum wave height. The greater the run-up height, the longer the waves will last when they rush inland, and the larger the area affected. The area affected will also be influenced by the relief of the land topography. Also important is the distance of an area from the source of a tsunami. Population density and early warning systems will play a role, as will the population resilience and quality of governance.

> This question is implicitly asking you to make a judgement ('the importance'), so you need to think about this as needing to create a rank order of factors, from most important to least.

> You need to think about both the physical and human factors that might come into force. Many students might concentrate on just the physical factors — something to watch out for.

Why do some tectonic hazards develop into disasters?

Tectonic hazards become tectonic disasters based on the balance and mix of a number of factors: magnitude, relative location and exposure or vulnerability of people. The level of development and the quality of governance contribute significantly to population vulnerability, and a lack of resilience can turn a hazard into a disaster.

Over the last 50 years, the total number of all hazards (not just tectonic) has increased, while the number of reported disasters and deaths has fallen. The economic costs of hazards and disasters of all types have risen considerably.

Therefore, a complex set of factors determines the effects of a tectonic disaster, both short- and longer-term. While magnitude may be important, it is often the location and characteristics of the local population that need consideration. This binds the ideas of risk, resilience and vulnerability.

Development and governance should never be underestimated in understanding disaster impact, vulnerability and resilience.

Practice questions ?

20 State the hazard–risk equation.

....

....

21 Define vulnerability **and** resilience in the context of hazards.

....

....

....

....

22 What is the difference between a hazard and a disaster?

....

....

23 What does VEI stand for, and how is it calculated?

....

....

....

24 List the characteristics commonly used in producing a tectonic hazard profile.

....

....

....

....

25 State **two** of the scales used to measure the magnitude and intensity of earthquakes.

....

....

26 Describe **two** social and **two** economic impacts of tectonic hazard events.

....

....

....

....

27 Study Table 1.1, which shows the number of people reported killed and affected by tectonic hazards between 2005 and 2015.

Table 1.1 People reported killed and affected by tectonic hazards between 2005 and 2015. Countries are grouped according to their level of human development. Red rows show those killed; orange rows show number of people affected

Tectonic risk	Level of human development (HD)				
	Very high	High	Medium	Low	Total
Earthquakes and tsunamis	21,036	38,019	293,941	297,328	650,324
Volcanic eruptions	0	21	330	12	363
Earthquakes and tsunamis	4,010,000	2,476,000	67,972,000	9,495,000	83,953,000
Volcanic eruptions	12,000	348,000	568,000	323,000	1,251,000

a Calculate the deaths from earthquakes and tsunamis, per year, in countries with very high levels of human development.

....

....

b Calculate the median deaths from volcanoes across the different levels of development.

Worked example

The data for number of people killed at each level of human development: 0, 21, 330, 12

Putting these into rank order: 0, 12, 21, 330

The median is the mid-point between 12 and 21: = 16.5

> This type of calculation question is quite common, so it is good to get confident in calculating measures of centrality: median, mean and interquartile range.

> Calculation of median gives you the middle value in a list of numbers. To find the median, your numbers have to be listed in numerical order from smallest to largest, so you may have to rewrite your list before you can find the median.

> The mean (average) of this set of data is 90.75. It is being pushed up by the very high 330 value. Which do you think is the better 'average', mean or median?

28 Suggest **one** additional measure of centrality that could be used to help interpret the data in Table 1.1. What are the different advantages and disadvantages of this technique?

....

....

....

....

29 Explain some of the ways in which disasters can create development opportunities for people and organisations.

....

....

....

....

....

....

30 Outline the reasons why low-income households and communities may be unfairly disadvantaged by an earthquake disaster.

....

....

....

....

....

....

31 Explain the advantages and disadvantages of using hazard profiles as a way of expressing tectonic risk.

....

....

....

....

32 Rank, in order of importance, what you consider to be the main economic impacts of tectonic hazards. Justify your choices.

....

....

....

....

....

....

33 Comment on the strength of the link between level of development and the capacity of a country to cope with a tectonic hazard.

....

....

....

....

....

....

34 Summarise and comment on the reasons why the impacts of earthquakes are generally greater than those of volcanoes.

....

....

....

....

....

....

How successful is the management of tectonic hazards and disasters?

Management of tectonic hazards often involves (1) improving the prediction of hazard events and (2) raising the general level of preparedness for hazards, for example through education and having rehearsed emergency procedures. Theoretical frameworks can be used to understand the predication, impact and management of tectonic hazards. Managing tectonic risk can be achieved by mitigation (sometimes referred to as adaptation or preparation). Improving technology allows the potential impacts of a hazard to be mitigated by making more effective adaptations, for example through aseismic building control. These vary in their effectiveness.

Practice questions ?

35 Define the term mitigation strategy.

....

....

36 What does the term NGO stand for?

....

37 Define the term multiple-hazard zone and give an example of a country or region that is exposed to multiple hazards.

....

....

....

38 Describe the different stages of the hazard management cycle.

....

....

....

....

39 What does the 'risk disk' model attempt to explain?

....

....

40 Describe an example of a local approach to protecting property from earthquake damage. State a place where it has been successfully used.

....

....

....

....

41 Give **two** examples of local protection from tsunamis.

....

....

42 Describe the different stages in Park's model of the disaster response curve.

....

....

....

....

43 Rank and rate the different priorities of the Sendai Framework for Action (2015).

....

....

....

....

Study Figure 1.2, which shows the number of hazard loss events between 1980 and 2018.

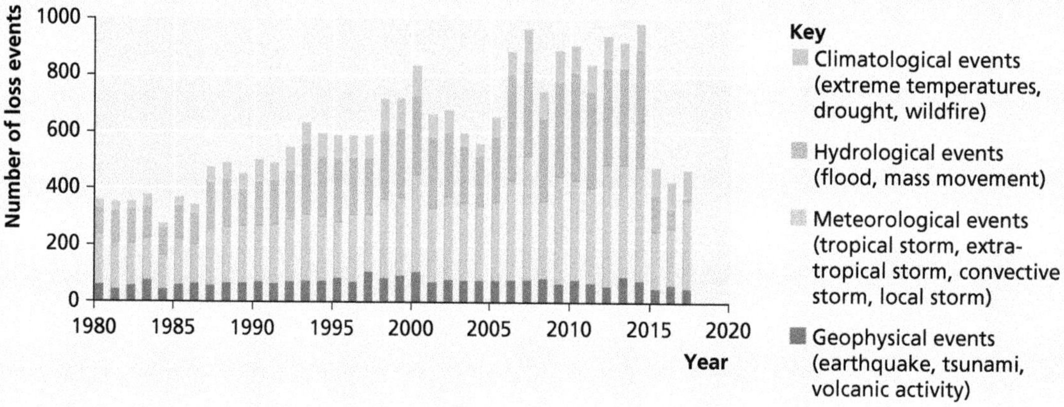

Number of loss events
1000
800
600
400
200
0
1980
1985
1990
1995
2000
2005
2010
2015
2020
Year
Key
Climatological events (extreme temperatures, drought, wildfire)
Hydrological events (flood, mass movement)
Meteorological events (tropical storm, extra-tropical storm, convective storm, local storm)
Geophysical events (earthquake, tsunami, volcanic activity)

Figure 1.2 Hazard loss events, 1980–2018

44 Describe the overall trend in the total number of hazard loss events in 1980–2018.

....

....

45 State **one** conclusion about the trend in geophysical (tectonic) hazards in relation to meteorological and hydrological hazards.

....

....

....

....

46 Explain the roles of different 'players' involved in managing the impacts of tectonic hazards.

....

....

....

....

....

47 Give reasons why disaster statistics may be inaccurate and/or have errors. What are the implications of this for people and organisations?

....

....

....

....

....

48 Explain the advantages and disadvantages of using a model like Park's in the study of tectonic disasters.

....

....

....

....

....

49 Explain the links between urbanisation, megacities and tectonic disaster risk.

....

....

....

....

....

....

1   a   Study Table 1.2, which shows data on the number of natural disasters and total damage for selected countries over a 10-year time period.

Table 1.2

Country	Number of natural disasters	Rank	Total damage ($ billion)	Rank	d	d²
China	286	1	265	2	–1	1
USA	212	2	443	1	+1	1
Philippines	181	3	16	7	–4	16
India	167	4	47	4	0	0
Indonesia	141	5	11	8	–3	9
Vietnam	73	6	7	9	–3	9
Afghanistan	72	7	0.16	10	...	...
Mexico	64	8	26	5	3	9
Japan	62	9	239	3	...	...
Pakistan	59	10	25	6	4	16

   i   Calculate the Spearman's rank correlation coefficient for the two sets of data by completing Table 1.2 and using the formula:

$$R_s = 1 - \frac{6\Sigma d^2}{(n^3 - n)}$$

where:

- R_s = is the Spearman's rank correlation coefficient
- n is the number of paired variables
- Σd^2 is the sum of the differences in rank squared

You must show your working. (AO3)   3 marks

⏱ 5

....

....

....

....

....

Table 1.3 shows the critical values for Spearman's rank *R* value and two hypotheses that are being tested.

Table 1.3

Confidence level	0.05 (95% significance)	0.01 (99% significance)
Critical value	0.564	0.746

Null hypothesis	There is no significant relationship between number of natural disasters (in a country) and total damage
Hypothesis	There is a significant relationship between number of natural disasters (in a country) and total damage

ii   Using the Spearman's rank correlation *R* value calculated in part i, state which hypothesis can be accepted. (AO3)   1 mark

....

....

b   Assess the extent to which the global distribution of tectonic hazards can be explained by plate boundary processes. (AO1, AO2)   12 marks

....

....

....

....

....

....

....

....

....

....

....

....

2   a   Study Table 1.4, which shows four different earthquakes and their impacts.

Table 1.4

Country	Year	Magnitude (MMS)	Death toll	Economic losses (US$ billions)
Italy	2009	6.3	8,900	16
Haiti	2010	7.8	316,000	8
New Zealand	2011	6.1	308	40
Nepal	2015	7.5	8,900	10

i   Calculate the mean number of deaths. (AO3)   1 mark

....

....

ii   Calculate the median MMS. (AO3)   1 mark

....

....

iii   Calculate the percentage difference between the highest and lowest economic losses. (AO3)   2 marks

....

....

b   Assess the role of governance in understanding the impacts of tectonic
    disasters in different countries. (AO1, AO2)   **12 marks**

....

....

....

....

....

....

....

....

....

....

....

....

....

3   a   Study Figure 1.3, which shows number of detected earthquakes between 2008 and 2018.

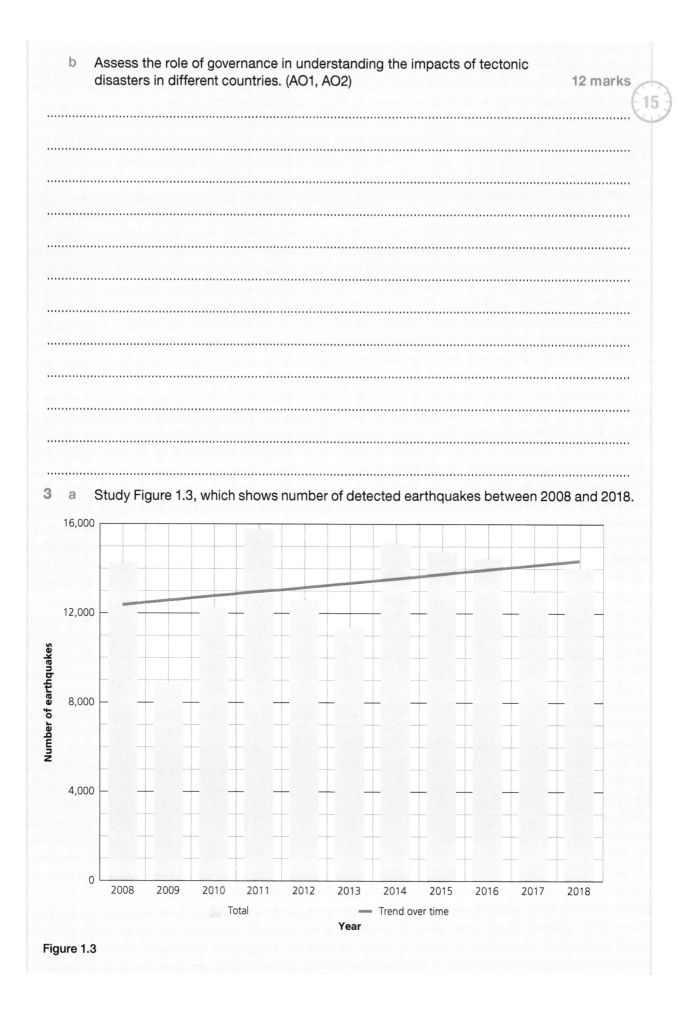

Figure 1.3

i   Calculate the range in the number of events in 2008–2018. (AO3)

1 mark

....

....

ii   Explain **one** reason why the use of the best-fit line (trend over time) may be inappropriate. (AO3)

3 marks

....

....

....

....

b   Assess how strategies to modify the event can reduce the impacts of tectonic hazards. (AO1, AO2)

12 marks

....

....

....

....

....

....

....

....

....

....

....

....

....

....

....

....

....

....

Topic 2A Glaciated landscapes and change

How has climate change influenced the formation of glaciated landscapes?

The Quaternary period in which we live is divided into two geological epochs: (1) the Pleistocene, in which there were multiple periods of warmer conditions (interglacials) as well as the cooler glacials; (2) the Holocene, which is the most recent interglacial. Students often forget that in this recent period too there have been many short-term fluctuations in climate. The causes of these temperature fluctuations include variations in the Earth's axis and in the composition of its atmosphere, as well as plate tectonics and sunspot activity.

The cryosphere consists of ice sheets and glaciers, together with sea and lake ice, ground ice (permafrost) and snow cover. Mass and energy are constantly being exchanged between the cryosphere and other major components of the Earth, namely the lithosphere, hydrosphere, atmosphere and biosphere. Variations in this cover have significant impacts on the Earth's albedo (reflectivity) and therefore form a direct link to climate change as well.

The term periglacial is now widely used to include all high-altitude or high-latitude areas with cold climates and land that may or may not have been glaciated. The particular conditions found in these areas lead to a unique set of processes and landforms.

Practice questions ?

1 What is meant by the term cryosphere?

....

....

2 What is the difference between an ice sheet and an ice cap?

....

....

....

3 What is mean by the term permafrost?

....

....

4 State **four** periglacial processes that are thought to occur.

....

....

....

....

5 Describe the key points of Milankovitch's theory.

....

....

....

6 What is meant by the term active layer?

....

....

....

7 What are **two** characteristics of warm-based glaciers?

....

....

....

....

8 What is the difference between a valley glacier and a piedmont glacier?

....

....

....

9 Compare the conditions of 'icehouse' and 'greenhouse' in relation to the Earth's climate.

....

....

....

....

10 Describe the process associated with the Earth's eccentricity.

....

....

....

11 Describe what is meant by a feedback mechanism in systems theory.

....

....

....

12 Describe **two** ways in which the global distribution of ice cover in the late Pleistocene differed from that today.

....

....

....

....

13 Explain **two** pieces of evidence for the Little Ice Age.

....

....

....

....

14 For **one** named periglacial landform, explain how it was formed.

....

....

....

....

15 Study Table 2.1, which shows estimates of present and past global ice cover.

Table 2.1

Region	Present area (estimated thousand km^2)	Past area (estimated thousand km^2)
Antarctica	1350	1450
Greenland	180	235
Arctic Basin	32	1600
Andes	3	88
European Alps	0.4	4
Scandinavia	0.4	660
Asia	12	390
Rest of the world	0.2	104
Total	1578	4531

a Calculate the mean reduction in ice cover for all regions.

....

....

b Calculate the percentage loss in Greenland.

....

....

c Identify the region with the biggest overall relative loss of ice.

....

16 Explain the factors limiting the global distribution of ice cover.

....

....

....

17 Explain the range of factors influencing the distribution and character of permafrost on a local scale.

....

....

....

18 Study Figure 2.1, which shows the climate at a periglacial location.

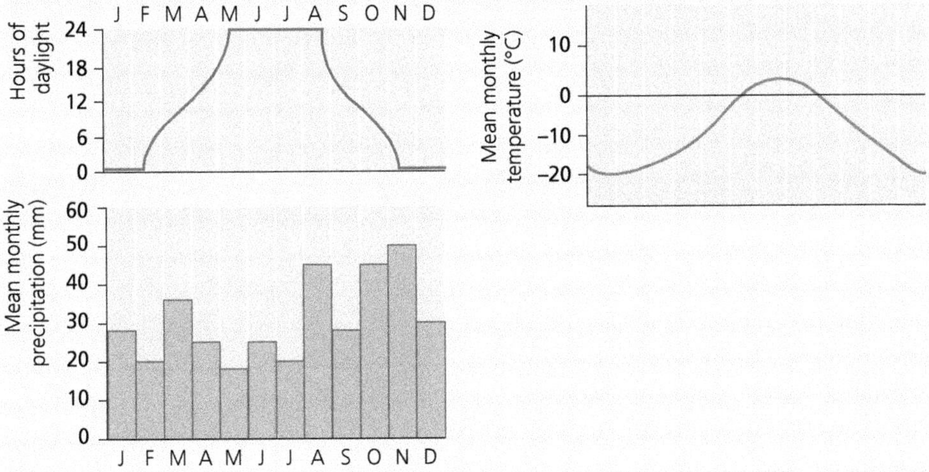

Figure 2.1

a Calculate the modal mean monthly precipitation.

....

....

b Calculate the number of months with a mean temperature below freezing.

....

....

c Describe the characteristics of the climate at this location.

....

....

....

....

d   Explain **one** reason why these data may be unreliable.

....

....

19 Explain the different shorter- and longer-term causes of climate change.

....

....

....

....

....

....

20 Explain the importance of different processes operating in periglacial environments.

....

....

....

....

....

....

....

What processes operate within glacier systems?

Systems theory is something that you are most likely familiar with from other parts of your geography A-level. Its relevance to both human and physical geography contexts, including glaciation, is no different, with inputs and outputs as well as interaction with other Earth systems. Mass balance relates to the relationship between the glacier's inputs (accumulation) and outputs (ablation). Accumulation tends to be greater than ablation in the upper part of a glacier, while the relationship is reversed in the lower part.

Gravity is obviously very important in the movement of ice and energy transfers. Ice moves downslope from higher altitudes either to lower locations on land or to sea level. The greater the ice mass (linked to accumulation of snow and ice), the more the conditions are set for ice flow. The rate of downslope movement also depends on the complex interplay of other factors.

The relationship between landscape and ice is complex. You need to remember that ice imprints on the land surface, but the significance of this depends on its scale, speed and load of debris. It is this debris that provides the grinding that is an essential part of erosion. The erosion takes place at the ice base and, in the case of glaciers, on the valley sides.

Study Figure 2.2, which shows a glacier mass balance.

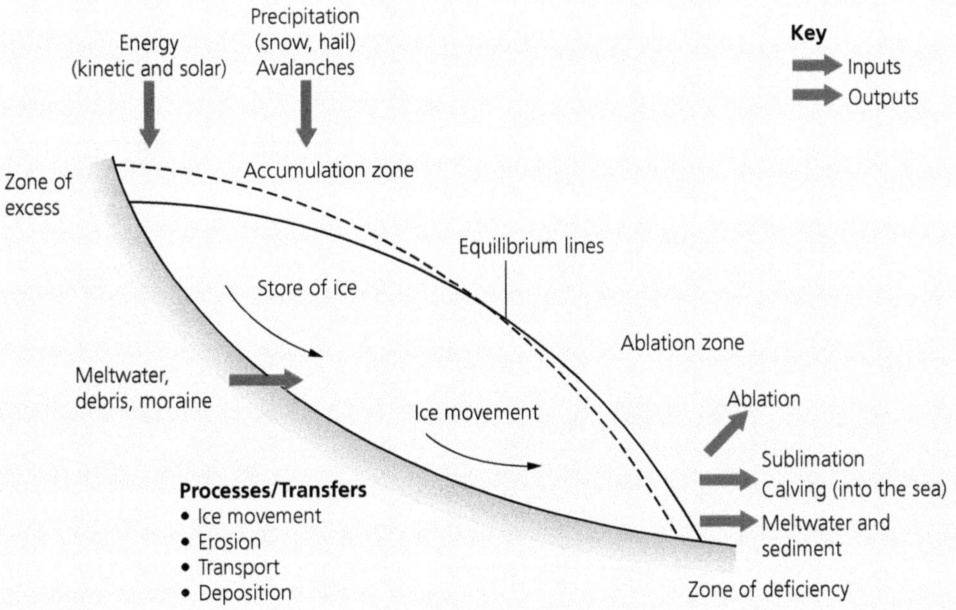

Figure 2.2

21 Identify **two** inputs to a glacier.

....

....

22 Identify **two** outputs from a glacier.

....

....

23 What is the difference between the terms englacial and proglacial?

....

....

....

24 What is meant by the term glacial surge?

....

....

....

25 What is meant by the equilibrium point of a glacier?

....

....

....

26　Describe **three** processes that are important to glacier movement.

....

....

....

....

27　Describe **one** way in which satellites and remote sensing are helping the study of ice sheets.

....

....

28　Complete the table to give **one** example of a landform at the three different scales.

Scale	Glacial landform example
Macro	
Meso	
Micro	

29　Describe the factors affecting the rate of glacier movement.

....

....

....

....

30　Explain the significance of a glacier's energy budget in the longer term.

....

....

....

....

31　Explain why polar and temperate glaciers move at different rates.

....

....

....

....

32　Explain how basal sliding and regelation flow operate.

....

....

....

....

33 Explain how negative feedback operates in the context of ice movement.

....

....

....

....

Study Figure 2.3, which shows a model of extending and compressing flow in ice.

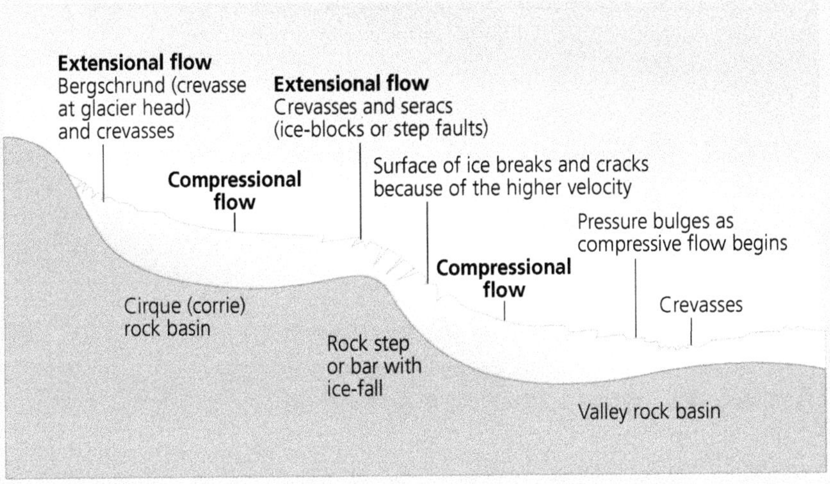

Extensional flow
Bergschrund (crevasse at glacier head) and crevasses

Extensional flow
Crevasses and seracs (ice-blocks or step faults)

Compressional flow

Surface of ice breaks and cracks because of the higher velocity

Pressure bulges as compressive flow begins

Compressional flow

Cirque (corrie) rock basin

Rock step or bar with ice-fall

Crevasses

Valley rock basin

Figure 2.3

34 Explain how changes in glacier flow are often linked to changes in valley gradient.

....

....

....

....

....

....

35 Explain how altitude and slope affect glacier movement.

....

....

....

....

....

....

36 Explain what is meant by the concept of a dynamic system. Why are glaciers included in this classification?

The physical landscape can be imagined as a series of linked components through which energy and material are cycled. Material is moved (transported) by geomorphological processes (e.g. mass movement or transport by ice) between stores. These systems are dynamic because they are constantly changing and reacting to different environmental conditions. A glacier can be considered as an open system, with inputs, storage, transfer and output of mass. It is a dynamic system, where the mass balance depends on the ratio of inputs and outputs.

> The key is to understand the idea of systems. This is one of the specialist concepts in the Pearson Edexcel specification. Remember that systems are composed of inputs, outputs, stores and transfers or flows. Systems will be important again when it comes to understanding carbon and water.

> This is a linking topic because you are drawing on knowledge and understanding that sit outside of your study of glaciation. It also means that you can direct your answer to what you feel confident writing about.

How do glacial processes contribute to the formation of glacial landforms and landscapes?

Glacial landforms are unique and remarkable. They can often be identified through their scale and impact on the landscape. It is all about the mass of ice and energy in the system. We already know that size and energy in the glacial system are strongly linked and have a direct effect on ice movement rate. Glacial mass also controls the efficiency of glacial erosion — its thickness and thermal regime. Glacial erosion operates more effectively when the glacier ice is warm-based and there is abundant meltwater and debris being transported.

The main processes of glacial deposition are lodgement, ablation, deformation and flow. Moraine is the term used to refer to an accumulation of glacial debris, whether it is dumped by an active glacier or left behind as a deposit after glacial retreat. Glacial deposition should not be confused with fluvioglacial deposition undertaken by glacial meltwater, which is a different process.

Glacial meltwater comes from surface melting and basal melting playing a significant role in the processes of erosion, transport and deposition. It is directly involved in glacier movement, providing the lubrication needed for basal sliding. It is also responsible for some erosion of the valley floor beneath the glacier, and is indirectly involved in the processes of glacial abrasion and plucking.

Practice questions ?

37 What is meant by the term stratification in relation to glacial deposits?

....

....

38 What is meant by the term ice-sheet scouring?

....

....

39 What is moraine?

....

....

40 Describe **three** glacial erosion processes.

....

....

....

....

....

41 Identify **three** types of glacial meltwater flow.

....

....

....

42 What is lodgement till?

....

....

43 What is meant by ice-contact depositional features?

....

....

44 Draw an annotated diagram to show the formation of a cirque.

45 Describe the differences between a terminal moraine and a recessional moraine.

....

....

....

....

46 Describe the landforms associated with glacial troughs.

....

....

....

....

47 Describe the different processes involved in the formation of drumlins.

....

....

....

....

48 Describe the ways in which fluvioglacial deposits differ from glacial deposits.

....

....

....

....

49 Explain the significance of the pressure melting point for ice movement.

....

....

....

....

50 Explain the roles of plucking and abrasion in the development of erosion landforms.

....

....

....

....

51 Explain the difference between ice-contact and proglacial fluvioglacial landforms.

....

....

....

....

52 Explain how the extent and direction of ice movement can be reconstructed from an examination of glacial depositional features.

....

....

....

....

....

....

53 Explain the circumstances under which a glacial meltwater stream might deposit its load.

....

....

....

....

....

54 Study Figure 2.4, showing the results of pebble orientation in a Scottish lowland glacial deposit.

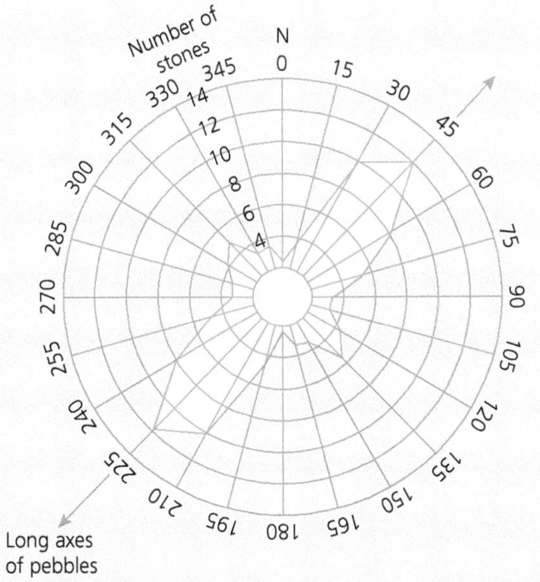

Long axes of pebbles

Figure 2.4

a Calculate the modal bearing.

....

....

....

b   Calculate the sample size.

Worked example

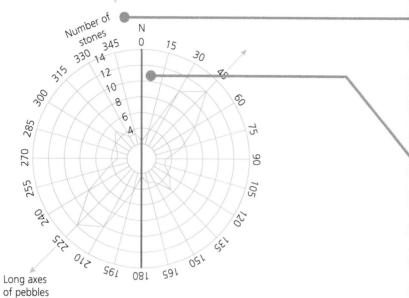

Rose diagrams are rather like a circular histogram — look at how frequency is labelled. You need to know this to tally the sample size, i.e. the number of stones in each category.

Remember that rose diagrams or polar graphs are presenting a mirror image of the data, so you do not want to double count. Here you just need to count the data on the right-hand side of the line.

Long axes of pebbles

Figure 2.5

Direction/bearing	Tally
0–14	2
15–29	10
30–44	12
45–59	8
60–74	5
75–89	3
90–104	3
105–119	4
120–134	5
135–149	3
150–164	3
165–179	2
Total	60

Orientation data have been collected every 15°, so you need to tally (total) up numbers from 12 sectors.

Total sample size = 60

c   Explain what the diagram indicates about the direction of ice movement.

....

....

....

....

....

....

55 With reference to specific landforms resulting from ice-sheet scouring, explain how differential geology is an important factor in their formation.

....

....

....

....

....

....

How are glaciated landscapes used and managed today?

In this particular part of the topic there are links to other parts of the specification, including water and carbon, as well as the more general themes around governance, culture and development. Be aware of the distinction between present and past glaciated landscapes. Present landscapes have more importance in this part of the topic. The active glacial and periglacial landscapes found today are most likely wilderness areas — places in the world remaining relatively untouched by human activity.

The main threats to these places come from human activity, although glaciated areas may be threatened by natural hazards. People often degrade landscapes as part of resource extraction, causing pollution and ecosystem damage.

Management of glacial landscapes is complex, involving lots of players and stakeholders. Possible approaches range from 'do nothing' or 'business as usual' to 'total protection'. Not all the players involved in glacial and periglacial environments share the same attitudes. What they want ranges from the outright exploitation of resources to total resource preservation.

Practice questions ?

56 What is meant by the term wilderness?

....

....

57 Identify **two** natural hazards that occur in glaciated areas.

....

....

58 Identify players whose actions are *directly* reducing the resilience of glaciated upland landscapes.

....

....

59 Identify players who are *indirectly* altering the natural systems of glacial and glaciated landscapes.

....

....

60 Describe **three** human threats to glaciated areas.

....

....

....

....

61 Describe contrasting attitudes that exist towards active glacial landscapes.

....

....

....

....

62 Describe Antarctica's unique system of international governance.

....

....

....

....

63 Describe the range of actions at a national level aimed at protecting glacial and glaciated landscapes.

....

....

....

....

64 Describe the potential economic value of glaciated areas.

....

....

....

....

65 Explain **two** impacts that global warming is having on glacier systems.

....

....

....

....

....

....

66 Study Figure 2.6, which shows some different opinions about the Antarctic Treaty System (ATS).

1 The ATS is one of the few international agreements of the 20th century to have succeeded.

7 Much of the science conducted in Antarctica is poor and is done to disguise territorial claims or potential rights to mineral exploitation.

8 There has been no armed conflict within Antarctica since the Antarctic Treaty was signed.

2 The ATS has maintained the spirit of peaceful international cooperation in Antarctica.

9 The ATS has focused only on the issues that are easily resolved, for example scientific cooperation, while avoiding fundamental problems such as competing territorial claims.

3 The ATS has limited environmental damage within Antarctica.

10 Antarctica is a 'common heritage for mankind' and should be governed as a 'world park' by the United Nations.

11 Government by consensus is a recipe for achieving the lowest common denominator at the slowest possible rate of progress.

4 The ATS has permitted Antarctic science to flourish and many issues of global concern, such as the ozone hole, have unfolded there.

12 The ATS has only succeeded because the principal Treaty nations feared what might happen if it failed.

5 The ATS has brought together many different nations, some of whom have been in conflict elsewhere in the world. For example, the USA and former USSR during the Cold War, and the UK and Argentina during the Falklands War.

13 The ATS doesn't provide any benefits to countries unable to pay for expensive scientific programmes within Antarctica.

6 The ATS is a 'rich man's club' run by a select group of developed countries for their own benefit.

Figure 2.6

a Giving reasons, which of the opinions do you:

i support most?

....

....

....

ii support least?

....

....

....

b Explain why some opinions might be unreliable.

....

....

....

....

....

67 Explain how direct actions by players increase the resilience of fragile glacial landscapes.

....

....

....

....

....

68 Explain why global warming is creating an uncertain future in glacial areas.

....

....

....

....

69 Explain the role of both mitigation and adaptation in protecting glacial wilderness areas and regions.

....

....

....

....

....

1 Study Figure 2.7, which shows an upland glacial landform in the UK.

Figure 2.7

a

i Explain how erosional processes have contributed to the formation of the feature shown. (AO1, AO2)

6 marks

8

....

....

....

....

....

....

ii Explain how subaerial processes have contributed to the development of this landscape feature. (AO1, AO2)

6 marks

8

....

....

....

....

....

....

b Explain why different approaches are needed to manage glaciated landscapes. (AO1)

8 marks

⑩

....

....

....

....

....

....

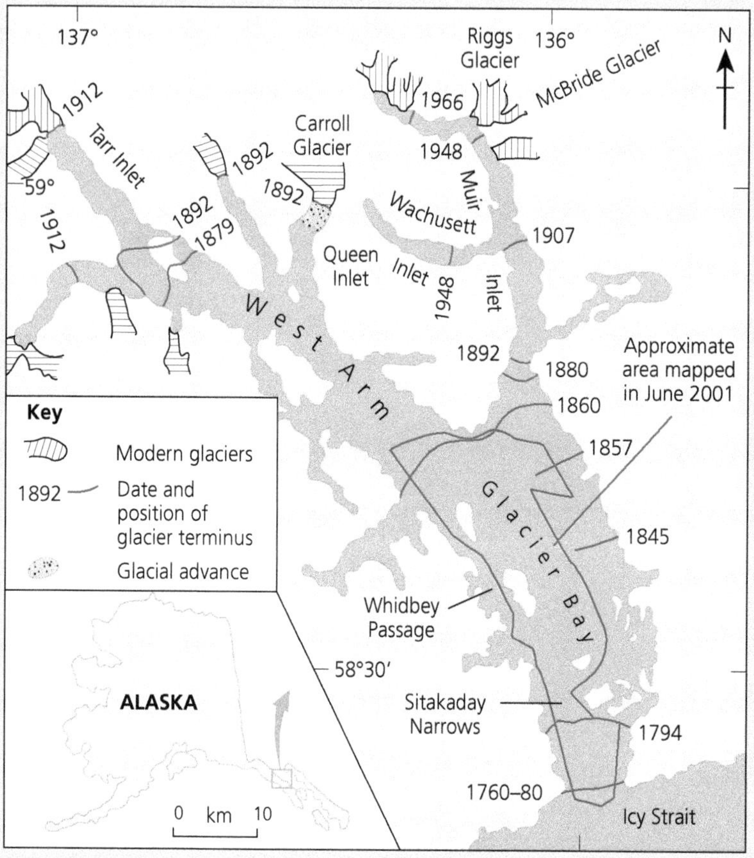

Figure 2.8

c Study Figure 2.8 above. Evaluate the view that climate change is critical in understanding differences in the rate of glacier movement. (AO1, AO2)

20 marks

㉕

....

....

....

....

....

2 Study Figure 2.9, which shows variations in the depth of permafrost.

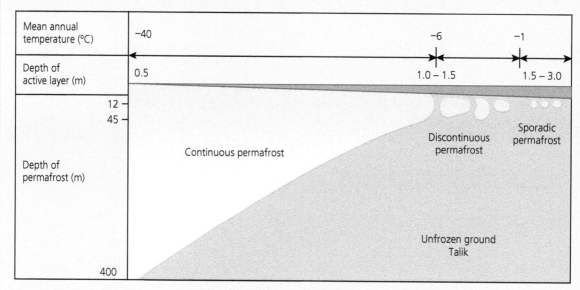

Figure 2.9

  a i Explain the variations in depth of the different types of permafrost.
(AO1, AO2)

6 marks

....

....

....

....

....

....

....

     ii Explain the link between Figure 2.9 and future climate change. (AO1, AO2) 6 marks

....

....

....

....

....

....

....

....

b   Explain how the glacial mass balance concept helps in an appreciation of the glacial system. (AO1)   8 marks

⟨10⟩

....

....

....

....

....

....

....

....

Study Figure 2.10, which shows the intensity of glacial erosion in Britain during part of the Pleistocene.

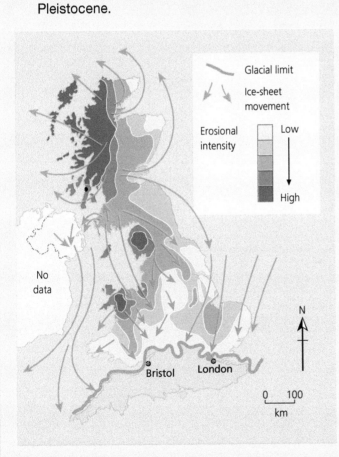

Figure 2.10

c   Evaluate the view that ice-sheet movement and erosional intensity can be accurately reconstructed from past ice events. (AO1, AO2)   20 marks

Write your answer on a separate sheet of paper and keep it with your workbook.

⟨25⟩

Topic 2B Coastal landscapes and change

Why are coastal landscapes different and what processes cause these differences?

The coast represents the boundary where land and sea meet, and where both marine and land processes interact with each other. The littoral zone consists of four subzones: the coast, the backshore, the nearshore and the offshore. This dynamic zone often changes rapidly, but contains a wide variety of coastal landscapes produced by the interaction of wind, waves and currents, as well as sediment from land and offshore sources.

Geological structure refers to three key ideas: strata (different layers of rock), deformation (degree of tilting or folding) and faulting (fractures causing rocks to be displaced from their original positions). Geological structure has a strong influence on coastal landscapes, often as much as, or more than, rock type or lithology.

The rate of coastal recession is controlled by several factors, including direction of coast, structure and lithology. Another significant factor is vegetation, particularly the degree of cover and general resistance to the forces of coastal erosion. Knowledge and understanding of the interaction of processes and conditions are critical in considering any scheme of coastal management.

Practice questions ❓

1   Describe **two** features of the backshore.

....

....

2   What is meant by the term Dalmatian coastline?

....

....

3   What is meant by the term subaerial processes?

....

....

4   Describe the key characteristics of: metamorphic, igneous and sedimentary rocks.

Metamorphic ....

....

Igneous ....

....

Sedimentary ....

....

5   What is meant by the terms concordant coast and a discordant coast?

....

....

....

....

6   What is a Haff coastline?

....

....

....

....

7   Describe some of the features of coastal plains.

....

....

....

....

8   What is the difference between a high-energy and a low-energy coastline?

....

....

....

....

9   Explain **two** ways in which geological structure influences rates of coastal erosion.

....

....

....

....

10 Explain how micro-features of a cliff are formed and characterised.

....

....

....

....

11 Explain **two** ways in which wind is important in the formation of sand dunes.

....

....

....

....

....

12 Explain **two** ways in which vegetation influences rates of coastal recession.

....

....

....

....

....

13 Complete the table about the rates of erosion of different rock types.

Rock type	Examples	Erosion rate	Explanation
Igneous			
Metamorphic			
Sedimentary			

14 Describe how the differential erosion of alternating and contrasting rocks affects the coastline.

....

....

....

....

....

15 Headlands and bays are a feature of discordant coasts. Explain why marine processes gradually smooth out such coasts.

....

....

....

....

....

16 Explain how dip influences cliff profiles.

....

....

....

....

....

17 Explain why estuaries are often associated with the development of salt marshes.

....

....

....

....

....

18 Explain why classification of coasts is rarely straightforward.

....

....

....

....

....

How do characteristic coastal landforms contribute to coastal landscapes?

Waves are an essential part of marine erosion, transport and deposition. They are caused by friction between the wind and the sea surface. The contact allows the wind's energy to be transferred into the sea. Any ripples created by this friction can be enlarged into waves if the wind is sustained or strengthens. Out at sea, waves are just energy moving through the water, with some orbital movement of water particles within each wave.

Material eroded from cliffs is transported by the sea as sediment in different ways. The movement of sediment is usually moved along the coast rather than at 90°. This transport is within larger sediment units — coastal sediment cells — which are understood as closed systems.

Weathering is important in the production of sediment, with mass movement transferring this material into the sea. These two processes are both significant factors affecting the rate of coastal recession. Both create distinctive landforms and, as a consequence, contribute to the appearance of coastal landscapes.

19 What is meant by the term hydraulic action?

....

....

20 What is meant by the term corrosion?

....

....

21 Suspension and solution are two processes by which sediment is transported. Name **two** other processes.

....

....

22 Name **three** types of weathering.

....

....

....

23 Name **three** types of mass movement.

....

....

....

24 Describe **two** differences between constructive waves and destructive waves.

....

....

....

....

25 Describe the process of longshore drift.

....

....

....

....

....

....

26 Coastal fieldwork was undertaken at Porlock beach in Somerset. A sample of 150 pebbles was collected as part of an investigation into beach processes. At each survey point the Cailleux roundness index was used to help understand stone shape. For this index a score between 0 and 1000 is calculated, where 0 is completely angular and 1000 is a perfect sphere. The results are shown in Table 2.2.

Table 2.2

Shape	Gore Point	Rank	Hurlstone Point
0–100	2	10	1
101–200	14	5	2
201–300	28	2	8
301–400	36	1	25
401–500	25	3	23
501–600	19	4	16
601–700	9	6	26
701–800	4	9	27
801–900	8	7	15
901–1000	5	8	7

a   Using Table 2.2 and Figure 2.11 complete **two** histograms, one for Gore Point and one for Hurlstone Point, to show the frequency of different pebble roundnesses at each end of Porlock beach.

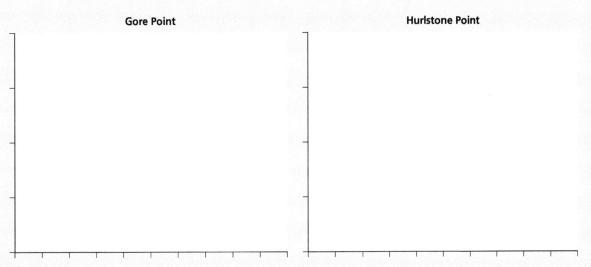

Gore Point

Hurlstone Point

Figure 2.11

b   What conclusions can be drawn from the histograms?

....

....

....

....

....

....

c   Describe another statistical technique that could be used to analyse these data.

Worked example

Spearman's rank could be used to compare the statistical relationships relating to sediment size between Gore Point and Hurlstone Point. Both sets of data would need to be ranked from highest to lowest (rank 1 for highest) and then the difference (d) for each rank calculated.

You should set up a table so that you can list the d values. These are then squared and then the squared values are summed to give Σd^2. This number is then fed into the formula to create an R_s value, which indicates the strength of the statistical relationship. Null and alternative hypotheses are used alongside confidence limits.

> The clue here is that there is a rank column included in the table, which should lead you to think about Spearman's rank. You could use other measures, for example the interquartile range or medians, to compare the two data sets.

> This is testing AO3 understanding, in this case using a variety of relevant quantitative, qualitative and fieldwork skills to interpret, analyse and evaluate data and evidence. In total AO3 accounts for 30% of marks, but much of it is found in paper 3 and in the independent investigation rather than in papers 1 or 2.

27 Describe what happens to waves when they reach shallow water.

....

....

....

28 Describe the formation of the following features:

    a   tombolo

....

....

    b   recurved spit

....

....

    c   cuspate foreland

....

....

29 Explain the differences between tides and currents.

....

....

....

....

30 Study Figure 2.12, which shows the relationship between climate and weathering.

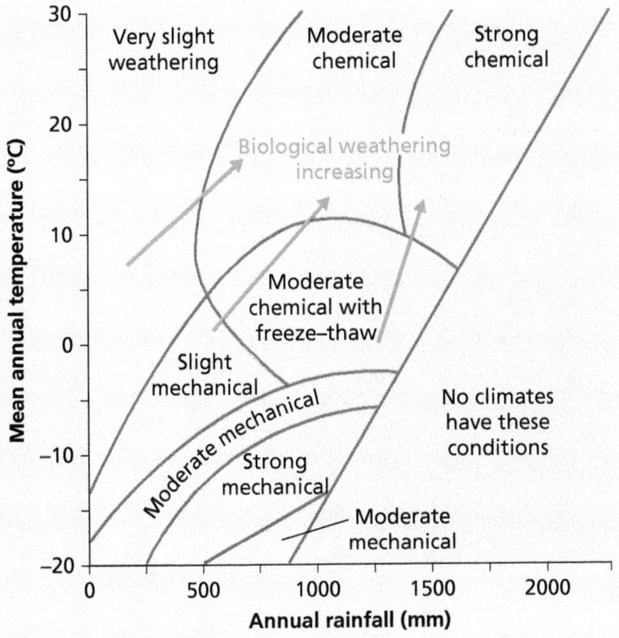

Figure 2.12

a   Explain the increase in biological weathering.

....

....

....

b   Compare cold and hot climates in terms of weathering.

....

....

....

c   Explain why in some parts of the world coastal landforms are significantly affected by subaerial processes.

....

....

....

....

....

....

31 Explain the different factors influencing the stability of depositional features.

....

....

....

....

....

....

....

32 Explain the link between subaerial processes and rates of coastal recession.

....

....

....

....

....

....

....

....

33 Explain the roles of both positive and negative feedback in the sediment cell concept.

....

....

....

....

....

....

....

....

....

....

How do coastal erosion and sea-level change alter the physical characteristics of coastlines and increase risks?

Changes in sea level influence the coast on different time scales, from daily, small-scale tidal movements to substantial shifts in mean sea level over thousands of years, as happened during the Ice Age. Sea-level rise linked to current global warming has, and will continue to have, a significant influence on the coastal landscape and its people.

Rising sea levels can pose a threat to people living in coastal zones, but this varies spatially. The challenge is made worse where there are large numbers of people living on or near the coast, with many of the world's largest cities occupying coastal locations. Linked to this is the economic generation of wealth, so it becomes a big challenge for governments and political establishments, which sometimes ignore sea-level changes or do not see them as a priority.

This set of ideas links to factors such as governance and habitat destruction, as well as natural processes, including storms. When certain physical and human factors come together you have a 'perfect storm' of potential devastation, where the risks are probably increasing in many places.

Practice questions ?

34 What is the difference between an isostatic change and a eustatic change in sea level?

....

....

....

35 Describe **one** cause of an isostatic change in sea level.

....

36 Name the type of coastline that results from a eustatic fall in sea level.

....

37 What is meant by the term storm surge?

....

....

38 Describe **two** coastal features created by a rising sea level.

....

....

....

39 Describe **two** ways in which technology is helping to measure coastal erosion.

....

....

....

40 Study Figure 2.13, which shows the features of an emergent coastline.

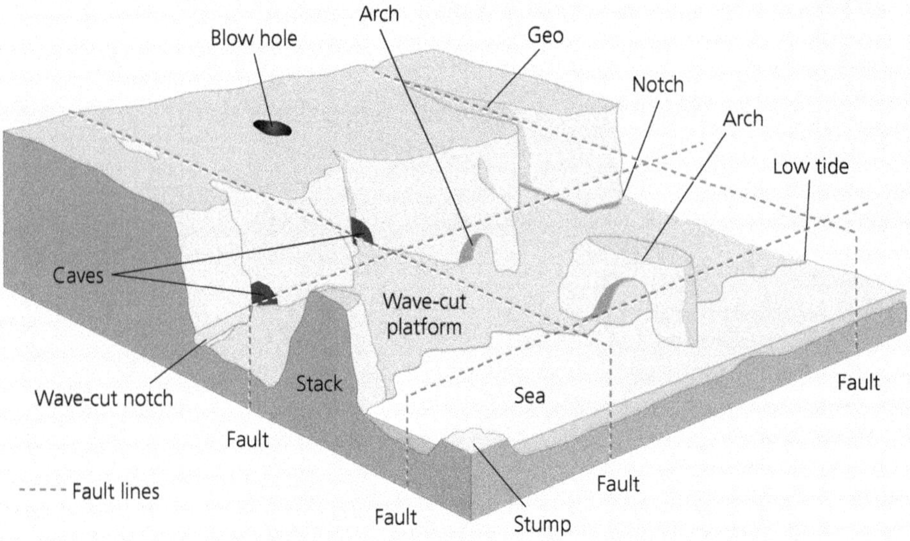

Figure 2.13

Using Figure 2.13, describe **two** reasons to suggest that this is an emergent coastline.

....

....

....

....

41 Identify **four** physical factors that encourage coastal recession.

....

....

....

....

42 Explain **one** reason why dredging can increase coastal erosion.

....

....

....

....

43 Explain the link between dams and coastal erosion.

....

....

....

....

44 Why do rates of coastal recession vary over time?

....

....

....

45 Study Figure 2.14, which shows past, present and forecasted global sea levels.

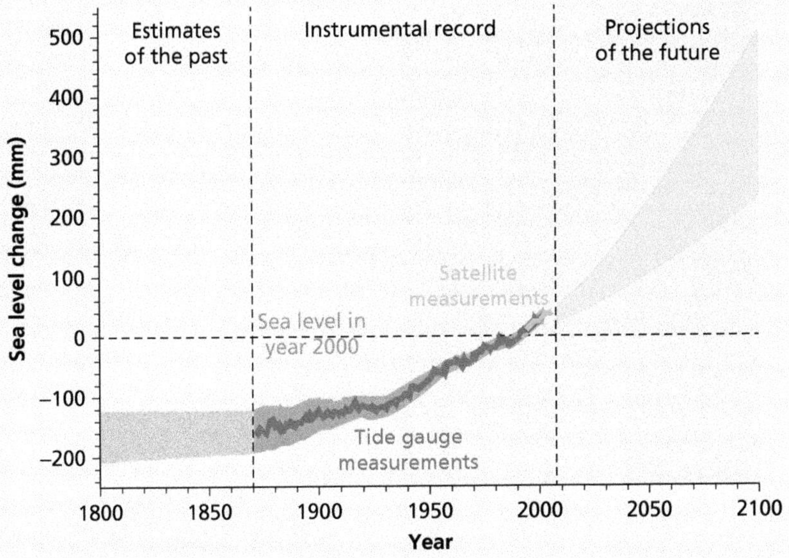

Figure 2.14

a Calculate the range of projected sea-level rise in 2100. Give the answer in mm.

....

....

....

....

b What do the estimates tell us about sea levels in the period 1800–70?

....

....

....

....

c Describe what the instrumental record tells us about the 20th century.

....

....

....

....

46 Explain how coastal topography has a direct link to the impact of storm surges.

....

....

....

....

47 Explain how tectonic activity can affect sea levels.

....

....

....

....

48 Explain the range of factors that make Bangladesh so vulnerable to severe coastal flooding and sea-level rise.

Worked example

Bangladesh is exceptionally vulnerable to climate change. Its low elevation, high population density and inadequate infrastructure all put the nation in harm's way, along with an economy that is heavily reliant on farming.

Two-thirds of Bangladesh is less than five metres above sea level, with an estimated 15 million people living in its low-lying coastal region. The process of salinisation has been exacerbated by rising sea levels. Coastal drinking water supplies have been contaminated with salt, leaving the 33 million people who rely on such resources vulnerable to health problems, such as pre-eclampsia during pregnancy, acute respiratory infections and skin diseases. Salinisation also leads to lower crop yields, creating a cycle of poverty.

> The key is to understand the geography of Bangladesh, including both the physical and human aspects. For example, you might think about governance, level of development or impacts from other hazards. This hazard also links to the high population density, inadequate infrastructure as well as the fact that Bangladesh is very low lying.

> This is a link question, since you are drawing on knowledge and understanding that sit outside of the topic of coastal landscapes. This means you can direct the answer to what you feel confident writing about.

49 Explain the different physical and human factors that make future sea-level rise difficult to predict.

....

....

....

....

....

....

How can coastlines be managed to meet the needs of all players?

The increased risks of coastal recession and coastal flooding threaten many communities with serious consequences, but this threat is uneven. The resulting losses tend to be localised, while the costs (economic, social and environmental) may have far-reaching consequences. In some parts of the world, the scale of recession and flooding is such as to soon require the wholesale abandonment of particularly low-lying islands and coastal zones. This may happen in parts of Bangladesh, for example.

Coastal management is probably something you are already familiar with. The classification of hard, soft and sustainable management is useful, but somewhat simplistic. While hard and soft engineering approaches are easily distinguishable, sustainable can encompass many different approaches. Generally, sustainable management aims to protect the people and their livelihoods in such a way as to minimise the ecological and environmental impacts of doing so. However, do not always conclude that this is the best option.

Modern coastal management follows a systems approach, often blending hard, soft and sustainable management in a 'holistic' framework. Coastal management is now planned on the basis of littoral cells, but in the end it is really money, political pressure and local campaigning that drive the types of management approach and their success.

Practice questions ?

50 What is meant by the term environmental refugees?

....

....

51 Describe what is meant by ICZM.

....

....

52 Describe what is meant by CBA.

....

....

53 Describe what is involved in a hard engineering approach to coastal management.

....

....

....

54 Describe the key features of ICZM.

....

....

....

....

55 Give **two** examples of economic and social losses resulting from coastal recession.

....

....

....

....

56 Describe what is involved in a soft engineering approach to coastal management.

....

....

....

57 Describe examples of assessments that would be carried out in an Environmental Impact Assessment (EIA) of a proposed sea defence project.

....

....

....

....

58 Give **two** reasons why some stakeholders/people object to hard engineering approaches.

....

....

....

59 Describe the difference between the terms mitigation and adaptation.

....

....

....

60 Explain why, in general, a soft engineering approach to coastal management is better than a hard engineering approach.

....

....

....

....

....

61 Explain the role and purpose of a Shoreline Management Plan (SMP).

....

....

....

....

62 Explain why coastal management is most likely to involve decisions that will divide
stakeholders into 'winners' and 'losers'.

....

....

....

....

63 Study Figure 2.15, which gives some land value data for England.

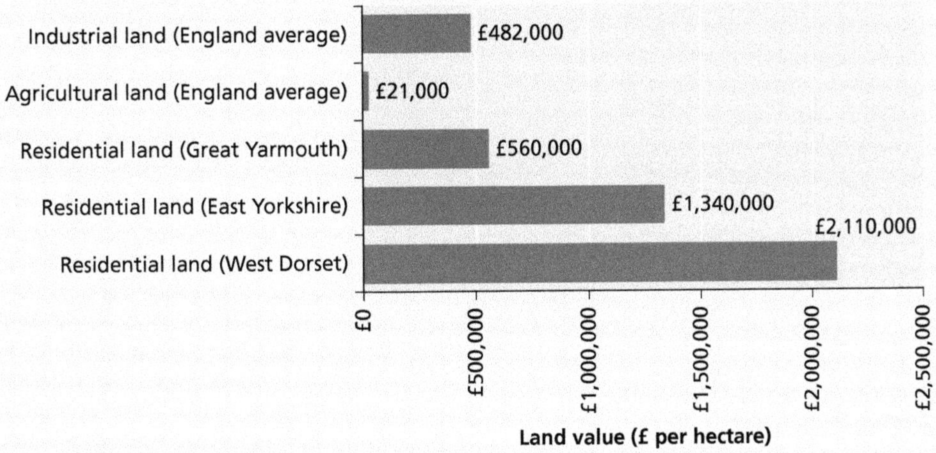

Figure 2.15

a   Calculate the mean land value per hectare. Give the answer to two decimal places.

....

....

b   Calculate the median land value per hectare. Give the answer to two decimal places.

....

....

c   Explain why there is a difference between the two values above. Suggest which is the
'best' measure, and why.

....

....

....

....

d   Explain the possible impacts on these land values where there is coastal recession.

....

....

....

....

....

....

64 Explain why the consequences of coastal flooding are generally greater in developing countries.

....

....

....

....

....

....

65 The risks associated with global warming are resulting in uncertainty in terms of coastal management. Explain why this is the case.

....

....

....

....

....

66 Explain how increasing globalisation affects the sustainable management of coastlines.

....

....

....

....

....

....

1 Study Figure 2.16, which shows a coastal landscape in North Norfolk.

Figure 2.16

a i Explain how erosional processes have contributed to a rapid rate of retreat
of this coastal landscape. (AO1, AO2)   6 marks

....

....

....

....

....

....

....

ii Explain how subaerial processes have contributed to the development of
this landscape. (AO1, AO2)   6 marks

....

....

....

....

....

....

....

b   Explain why coastal landscapes have economic value. (AO1)   8 marks

10

....

....

....

....

....

....

....

....

....

Study Figure 2.17.

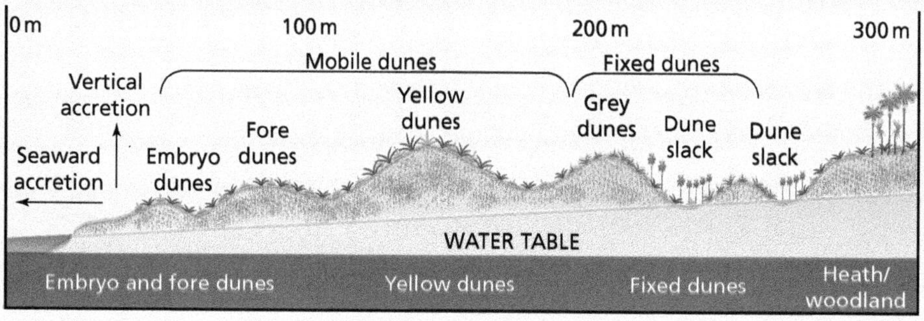

Figure 2.17

c   Evaluate the view that coastal ecosystems such as sand dunes provide
     better flood protection compared with hard defences. (AO1, AO2)   20 marks

20

....

....

....

....

....

....

....

....

....

....

2 Study Figure 2.18, which shows a sketch map of the beaches around Newquay, Cornwall.

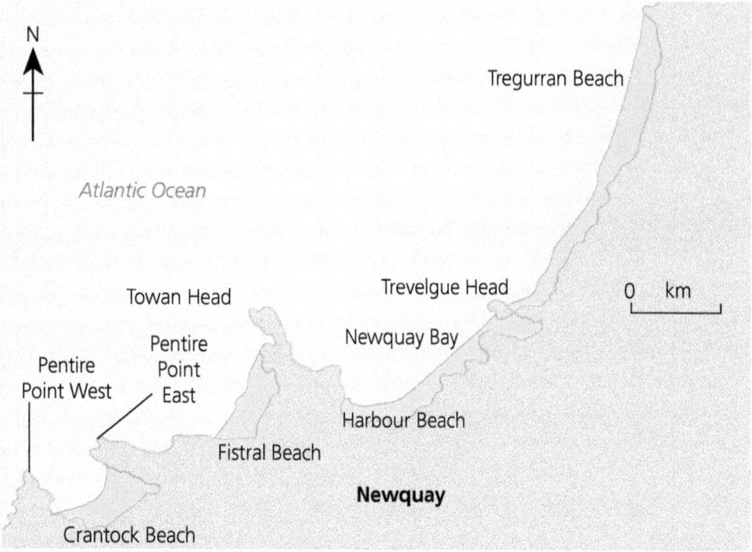

Figure 2.18

a i Explain how deposition and erosion have played a role in shaping this part of the coast. (AO1, AO2)

6 marks

....

....

....

....

....

....

....

ii Explain the different impacts that wind direction may have on processes operating within the area shown by the map. (AO1, AO2)

6 marks

....

....

....

....

....

....

....

b Explain why there are conflicts between stakeholders when it comes to coastal management. (AO1)

8 marks

10

....

....

....

....

....

....

....

....

....

Study Figure 2.19, which shows past, present and forecasted global sea levels.

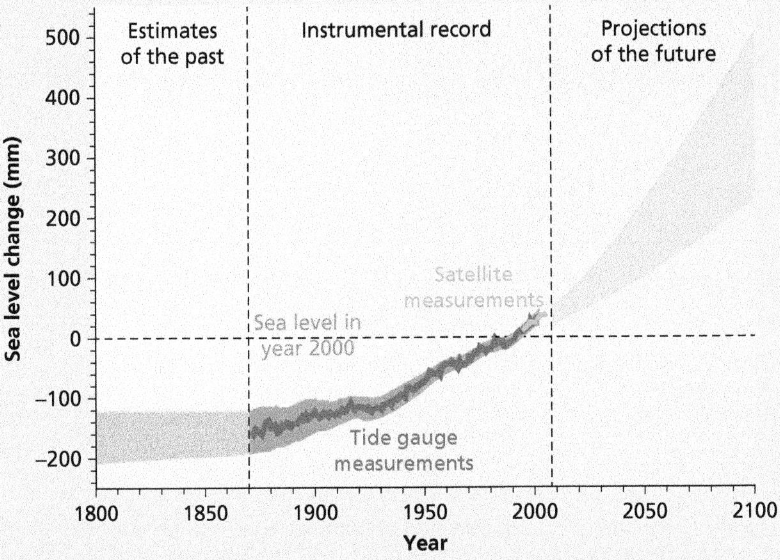

Figure 2.19

c Evaluate the view that climate change is the most important factor in influencing coastal flood risk. (AO1, AO2)

20 marks

20

....

....

....

....

....

Workbook answers at **www.hoddereducation.co.uk/workbookanswers**

Topic 5 The water cycle and water insecurity

What are the processes operating within the hydrological cycle?

Much of the geography in this section is very familiar, although it is tied together into a more complex systems-based approach. This is something you will have already covered in other parts of this workbook.

The global hydrological cycle is a closed system, involving stores, flows and processes that are driven by solar and gravitational energy. The stores differ in their relative importance and size. The global water budget determines the amount of water available for human use.

The drainage basin is an open system within the global hydrological cycle. Physical factors within a drainage basin determine its inputs, flows and outputs. However, human activities disrupt the drainage basin cycle by accelerating processes, creating new stores and abstracting water.

The hydrological cycle influences water budgets and river systems at a local scale. Water budgets are determined by climate and have an impact on water availability. River regimes, too, are influenced by climate, but also by geology and soils. Storm hydrographs are an important aspect of the behaviour of river regimes.

Practice questions ?

1 What is meant by the terms store and flow?

....

....

2 What does the following equation mean?

$P = R + E \pm S$

....

3 List the main inputs of a drainage basin system.

....

....

....

4 What is the difference between interception and infiltration?

....

....

....

5   What is the difference between blue water and green water?

....

....

....

....

6   Give an example of positive feedback in the water cycle.

....

....

....

7   Draw a simple storm hydrograph and label it.

8   What is the difference between a water budget and a river regime?

....

....

....

....

9   Identify **three** ways in which people disrupt the drainage basin cycle.

....

....

....

....

10 Study Figure 5.1.

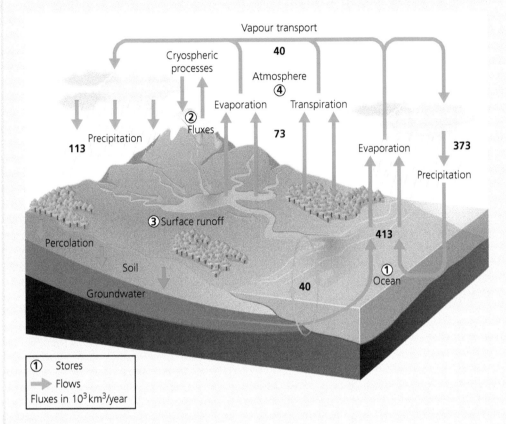

Figure 5.1

a Calculate the total flows from precipitation.

....

b Describe **two** problems with this diagram.

Worked example

Firstly, the diagram does not have values for all the different flows, for example evaporation and transpiration. There are also no values for percolation, soil flows, etc.

A second problem is that the stores are identified but there is no indication of their relative size or scale. This means that making judgements about their importance, for example, is difficult to do since we cannot see their spatial extent.

A third issue is that this model is generalised and figures need to be seen as 'best guesses' rather than absolute data, since at this scale the flows and stores can only be modelled and estimated rather than actually measured to give empirical data.

> There is no indication of the relative sizes of stores, so making judgements about scale or importance of these different components is difficult.

> The global water cycle is a model of how scientists expect the system to work. It is very much a simplification of reality, with numbers provided as a rough guide. This is useful knowledge if you are ever asked to assess, judge or analyse information such as this.

11 What are the **three** largest stores in size order?

....

....

12 Describe how geology affects the flows in a drainage basin.

....

....

....

13 Describe how deforestation is affecting the Amazon basin.

....

....

....

14 Explain why the global hydrological cycle is a closed system.

....

....

....

15 Explain why the drainage basin is classified as an open system.

....

....

....

16 Explain why only a small percentage of freshwater is accessible for human use.

....

....

....

....

17 Study Figure 5.2, which shows runoff in the drainage basins of three rivers in Western Europe.

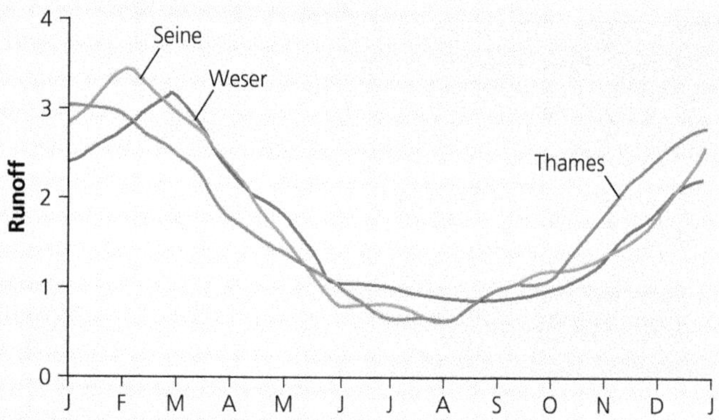

Figure 5.2

Explain the reasons for the seasonal variations in runoff (assuming rainfall is evenly distributed).

....

....

....

....

18 Explain how different types of geology affect the shape of a storm hydrograph.

....

....

....

....

19 Explain how temperature, humidity and wind affect evapotranspiration.

....

....

....

....

....

....

....

20 Explain how climate affects water budgets.

....

....

....

....

....

....

21 Explain how rising urbanisation increases the risk of flooding.

....

....

....

....

....

What factors affect the hydrological cycle?

The hydrological cycle is a familiar concept in geography. This area of the specification links water budgets and river systems at different scales. Aspects covered include droughts in the context of water deficits. Although the root causes of droughts are meteorological, their impacts fall under the two broad headings of hydrological and agricultural. Droughts are often associated with famines, and can have several synoptic links in terms of other hazards, wealth and quality of life.

Flooding tends to hit the headlines more often than droughts. While the root causes of flooding, as with droughts, are meteorological, other physical factors play their part. These include basin-specific factors, such as geology, soil, topography and vegetation. But the flood risk can be exacerbated by a range of human actions. Risk and resilience are recurring themes included in this topic.

Another synoptic theme is climate change and its link to the hydrological cycle. Global warming is going to intensify and accelerate the cycle. However, the impacts will be uneven and spatially varied.

Practice questions

22 What does ENSO stand for?

....

23 Identify **three** meteorological causes of flooding.

....

....

....

24 Study Figure 5.3, which shows rainfall trends in the Sahel region.

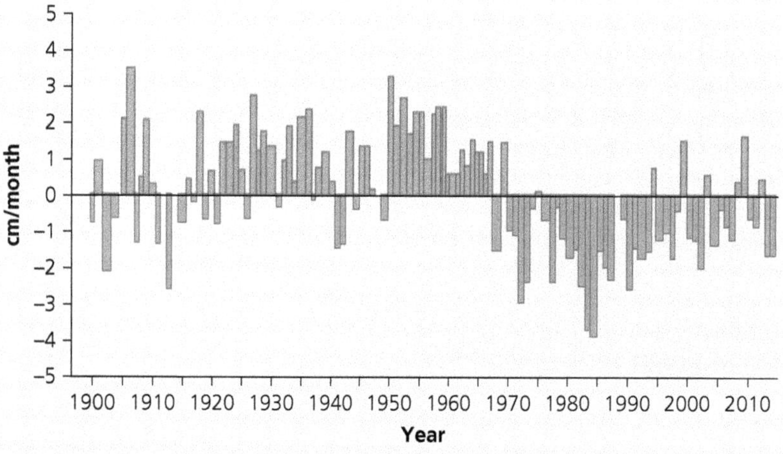

Figure 5.3

a Calculate the range in rainfall (cm/month).

....

....

....

b　Explain **one** reason why these data may be unreliable.

....

....

c　Describe reasons for the changing trends in rainfall since 1900.

....

....

....

25　Describe how short-term climate change threatens to diminish water supplies.

....

....

....

26　Describe **three** ways in which people are contributing to desertification.

....

....

....

27　Describe the different impacts flooding can have on ecosystems.

....

....

....

28　Explain **one** way in which hard engineering can contribute to flooding.

....

....

....

....

29 Study Figure 5.4, which shows the human factors affecting a flood hydrograph.

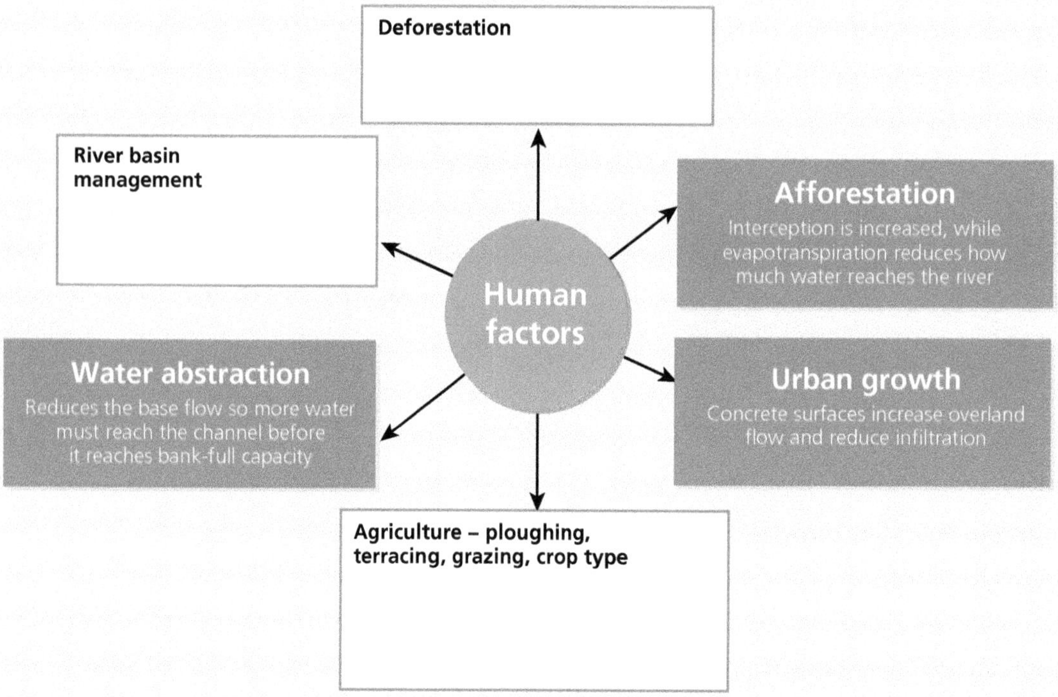

Deforestation		

River basin management

Human factors

Afforestation
Interception is increased, while evapotranspiration reduces how much water reaches the river

Water abstraction
Reduces the base flow so more water must reach the channel before it reaches bank-full capacity

Urban growth
Concrete surfaces increase overland flow and reduce infiltration

Agriculture – ploughing, terracing, grazing, crop type

Figure 5.4

Complete the blank boxes to explain how each factor will affect a river's response to a storm.

30 Describe how the inputs and outputs of the global hydrological cycle might be changed by global warming.

....

....

....

....

....

31 Drought is described as a 'creeping hazard'. Explain what this idea means.

....

....

....

....

....

....

32 Explain the different human impacts as a result of flooding.

....

....

....

....

....

....

33 To what extent are Australia's droughts natural or provoked by people?

....

....

....

....

....

....

34 Explain the wetland functions that are threatened by drought.

....

....

....

....

....

....

35 Explain the link between more water circulating around the global hydrological cycle and more energy in the atmosphere.

....

....

....

....

....

....

36 Explain why the forecasting of possible global warming is full of uncertainties.

....

....

....

....

....

....

How does water insecurity occur and why is it becoming such a global issue for the 21st century?

An important global issue, which has social, economic, environmental and political ramifications, is the growing mismatch between water supply and water demand. This is creating global patterns of water stress, scarcity and insecurity. These are mainly the result of human rather than physical factors, such as population increases and development. These stress precursors overlap with the carbon and energy topics.

There are consequences and risks associated with water insecurity. Access to water has the potential to become a source of conflict between countries — between the 'haves' and the 'have-nots'. Other potential flashpoints are international rivers and other trans-boundary water sources.

A hard engineering approach to water management is complex and costly. Increasingly more sustainable water schemes and a more comprehensive approach to the management of large river basins are being utilised. International agreements about water sharing are also being put in place, but not always effectively.

Practice questions ?

37 What is meant by the term water insecurity?

....

....

38 What is the threshold below which water shortages are said to exist?

....

....

39 What is meant by the environmental sustainability of water?

....

....

40 What is the aim of the Helsinki Rules?

....

....

....

41 Describe **three** different ways of conserving water.

....

....

....

42 What is meant by saltwater encroachment, and where does it occur?

....

....

....

43 Describe **three** factors affecting the price of water.

....

....

....

44 Give examples of the main players involved in issues relating to water resources.

....

....

....

45 Describe the environmental problems resulting from an inadequate water supply.

....

....

....

46 Describe how the El Niño–Southern Oscillation (ENSO) can be linked to water conservation challenges.

....

....

....

....

47 Describe what is meant by integrated water resource management (IWRM).

....

....

....

....

48 Table 5.1 shows changes in water stress for selected countries, 1992–2017 (%).

Table 5.1

	1992	1997	2002	2007	2012	2017
China	36.5	38.4	40.2	41.7	43.8	43.4
France	34.3	26.6	28.3	27.1	24.2	23.2
Israel	136.7	132.2	133.1	122.4	115.6	103.4
Tunisia	77.9	72.3	71.3	N/A	81.7	121.1

Source: http://www.fao.org/nr/water/aquastat/data/query/results.html

Note that water stress (%) is calculated as:

$$\frac{\text{total freshwater withdrawn}}{\text{total renewable freshwater resources} - \text{environmental flow requirements}} \times 100$$

a Use Figure 5.5 to plot the data for the four countries.

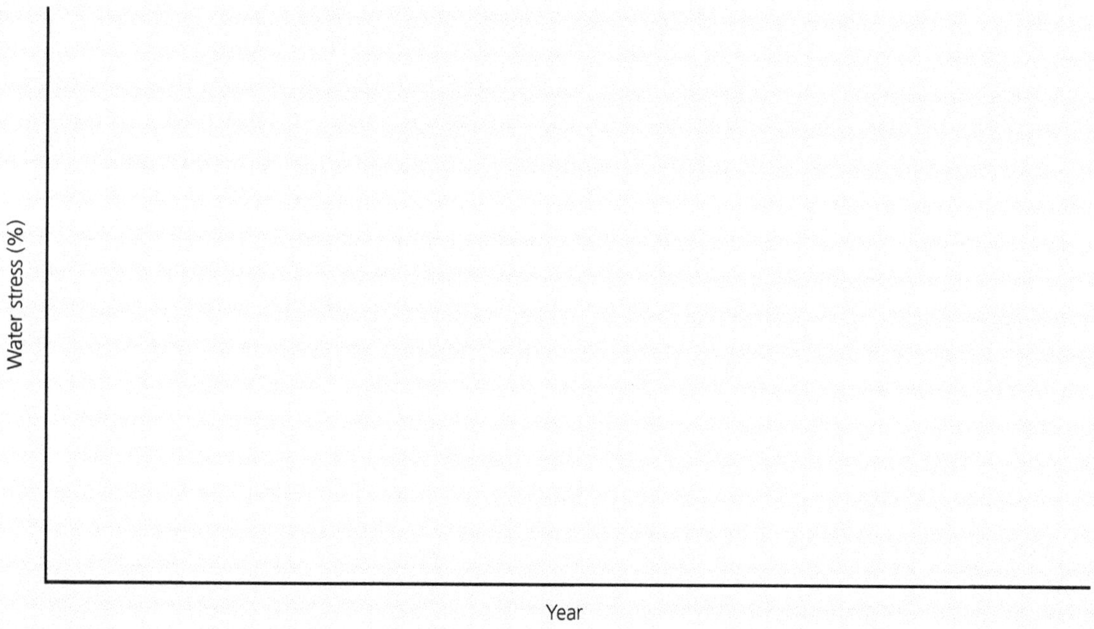

Water stress (%)

Year

Figure 5.5

b Calculate the percentage change in water stress in Israel between 1992 and 1997. Give the answer to two decimal places.

....

....

....

c In which country and in which 5-year time period was there the greatest relative change in water usage? Calculate your answer.

Worked example

$$\text{relative change (\%)} = \frac{\text{final value} - \text{initial value}}{\text{initial value}} \times 100$$

Absolute and relative changes are important ideas. The absolute difference is the final value minus the initial value. So, for Tunisia, the absolute change between 2012 and 2017 was 121.1 − 81.7 = 39.4.

Tunisia appears to have the largest relative change, in 2012–2017:

$$\text{final value} = 121.1$$
$$\text{initial value} = 81.7$$
$$\frac{121.1 - 81.7}{81.7 \times 100} = \textbf{48.2\%}$$

It is possible that the smallest and biggest absolute and relative changes are in fact the same data points. For example, when a value of 1 doubles, the absolute increase is only 1, but the relative change is 100%.

Relative difference is a little more complex. The relative change is a fraction that describes the size of the absolute change in comparison with the reference value. In a table of data like this it is best to look for what appears the biggest relative change or 'jump'. Tunisia 2012–2017 is clearly large. All other increments are small — based just on a visual assessment.

d Explain **one** problem with the data in Table 5.1.

....

....

....

....

49 Explain how the water poverty index is calculated.

....

....

....

....

....

50 Describe how there may be conflict between water users in the same country.

....

....

....

....

....

....

51 Explain the main physical causes of water insecurity.

....

....

....

....

....

....

52 Explain the link between development and increasing demand for water.

....

....

....

....

....

....

53 Explain the risks associated with water transfer schemes.

....

....

....

....

....

....

54 Explain why it is difficult to make accurate forecasts of future water scarcity.

....

....

....

....

....

....

1   Study Figure 5.6, which shows the converging costs of supplying fresh and desalinated water in the USA.

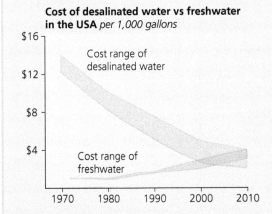

Cost of desalinated water vs freshwater in the USA *per 1,000 gallons*

Figure 5.6

a   Explain **one** reason why the two cost curves are converging. (AO1, AO2)   3 marks

4

....

....

....

b   Explain why land use changes can increase flood risk. (AO1)   6 marks

8

....

....

....

....

....

....

c   Explain why, at a global scale, the price of water can vary in different places. (AO1)   8 marks

10

....

....

....

....

....

....

....

Study Figure 5.7, which shows the global distribution of water scarcity.

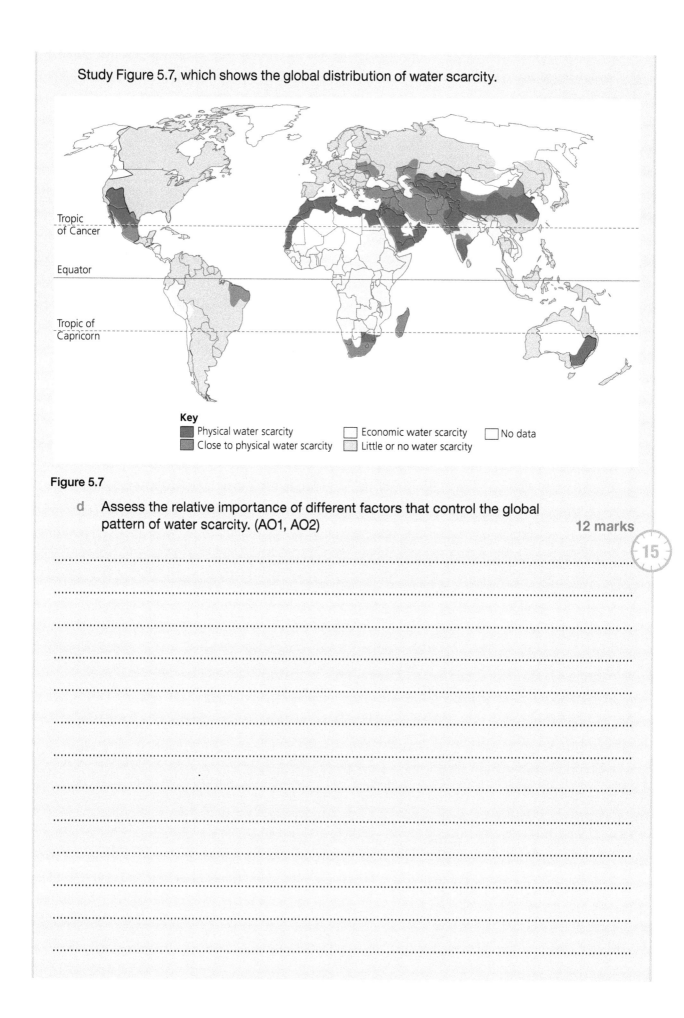

Key
■ Physical water scarcity
■ Close to physical water scarcity
□ Economic water scarcity
□ Little or no water scarcity
□ No data

Figure 5.7

d Assess the relative importance of different factors that control the global pattern of water scarcity. (AO1, AO2)

12 marks

(15)

....

....

....

....

....

....

....

....

....

....

....

....

....

e Evaluate the view that large-scale water management projects often create more challenges than opportunities. (AO1, AO2)

20 marks

25

2   Study Figure 5.8, which shows the trends in sectoral water use.

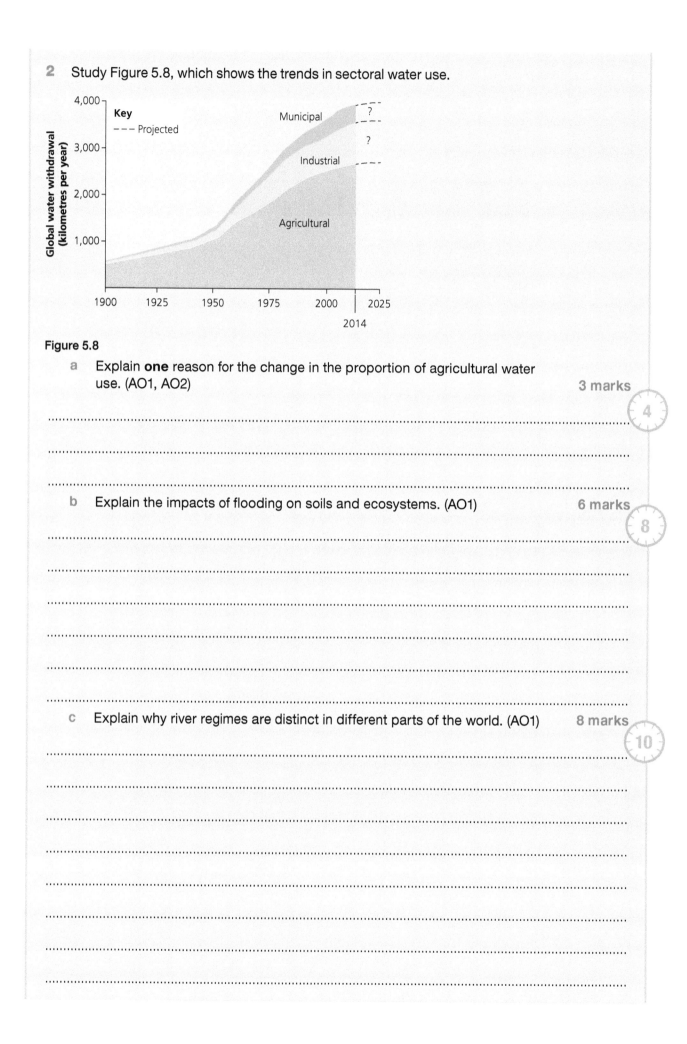

Figure 5.8

a   Explain **one** reason for the change in the proportion of agricultural water use. (AO1, AO2)   3 marks

....

....

....

b   Explain the impacts of flooding on soils and ecosystems. (AO1)   6 marks

....

....

....

....

....

....

c   Explain why river regimes are distinct in different parts of the world. (AO1)   8 marks

....

....

....

....

....

....

....

....

Study Figure 5.9, which shows some of the challenges associated with balancing water transfer schemes.

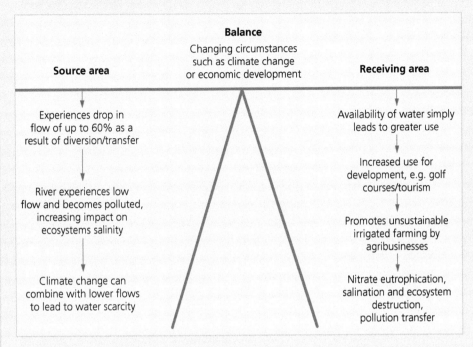

Figure 5.9

d   Assess the relative importance of different issues associated with water transfer schemes. (AO1, AO2)

12 marks

15

....

....

....

....

....

....

....

....

....

....

....

....

....

....

....

....

e   Evaluate the view that people are the most important factor causing drought
     to occur. (AO1, AO2)   20 marks

     25

....

....

....

....

....

....

....

....

....

....

....

....

....

....

....

....

....

....

....

....

....

....

Topic 6 The carbon cycle and energy security

How does the carbon cycle operate to maintain planetary health?

The main features of the carbon cycle should be familiar to you from GCSE science. For A-level geography there will be several areas with new terminology, which makes it potentially complex and a barrier to understanding, so compiling a glossary is recommended. Try to think about the relative sizes of stores and flows, and build up a mental map to help you evaluate information when required. This topic includes many complex social, economic and geopolitical links to other aspects of geography.

The carbon cycle operates at different spatial and temporal scales. Different physical processes control the movement of carbon between stores on land, in the oceans and in the atmosphere. Biological processes sequester carbon both in the oceans and on land, tending to operate as a 'fast' cycle. Phytoplankton and terrestrial primary producers sequester atmospheric carbon during photosynthesis. Biological carbon can be stored as dead organic matter either on the ocean floor or in the soil.

A balanced carbon cycle is important. However, the burning of fossil fuels is a persistent and controversial issue, with little global agreement on how to deal with the problems it brings.

Practice questions ?

1 List **three** rocks that store carbon.

....

....

....

2 Describe the difference between a store and a flux.

....

....

....

3 Name two factors controlling the movement of carbon within the oceans.

....

....

4 What is meant by the process of sequestration?

....

....

5   What is biological carbon?

....

....

....

6   Describe how terrestrial carbon is returned to the atmosphere.

....

....

....

7   Describe the terrestrial part of the carbon cycle.

....

....

....

8   Describe the role of geological fluxes in the carbon cycle.

....

....

....

9   Describe the role of thermohaline circulation.

....

....

....

10   Describe the main processes involved in photosynthesis.

....

....

....

11   Explain the process of photosynthesis.

....

....

....

....

12 Study Table 6.1, which shows a sample of carbon fluxes.

Table 6.1

Flux	Rate (PgC per year)
Photosynthesis	103
Respiration	50
Volcanic eruption gases	50
Diffusion from ocean	9.3
Vegetation to soil decomposition	5
Weathering and erosion	0.9
Sedimentation/fossilisation	0.2

a Calculate the percentage difference in fluxes between weathering and erosion and photosynthesis.

....

b If these data were plotted as a bar chart, explain **one** difficulty with completing the plot.

....

....

....

Worked example

There is a very large range of rates in this table of carbon fluxes — from 0.2 to 103. This would make the plotting of any bars problematic because the lowest values (0.5–5 PgC/yr) would be so small as to be unreadable against a scale that went from 0 to 105 on a piece of linear graph paper.

Logarithmic paper can be used when there is a wide range of values on a graph. You sometimes see graphs plotted as log graphs (along the x-, y- or both axes) as part of A-level geography exams.

13 Complete Table 6.2 to show the four main stores of carbon, and state the form of carbon present in each store. An example has been completed for you.

Table 6.2

Name of carbon store	Form of carbon present
Ocean	Dissolved inorganic carbon and carbon dioxide

14 Explain the link between carbon sequestration and photosynthesis.

....

....

....

....

15 Explain **one** process, other than burning fossil fuels, that people use to balance the carbon cycle.

....

....

....

....

16 Explain the different factors that determine the capacity of a soil to store organic carbon.

....

....

....

....

....

....

17 Study Figure 6.1, which shows the oceanic carbon pumps.

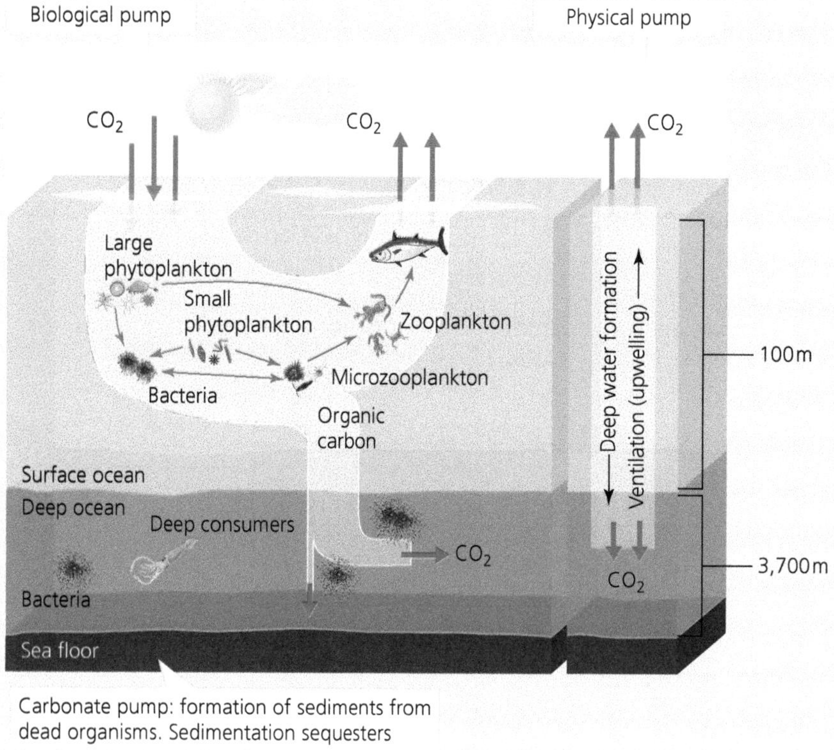

Figure 6.1

Explain the role of these pumps in the carbon cycle.

....

....

....

....

....

....

....

18 Explain the link between the burning of fossil fuels and the greenhouse effect.

....

....

....

....

....

....

....

....

What are the consequences of the increasing demand for energy?

Management of energy is one of the biggest challenges people and governments face in the 21st century. It is a synoptic topic because players control, abuse and dominate complex economic systems. Achieving energy security is a key goal for all countries, particularly those relying heavily on fossil fuels to meet their energy needs. Access to, and consumption of, energy depend on the physical availability of resources, cost, technology, level of economic development and environmental considerations.

There is a mismatch between the distribution of fossil fuel resources and the places where demand is highest. This makes energy an economic and geopolitical weapon. Energy pathways provide the links, but are complex and prone to disruption. The development of unconventional sources of fossil fuels has caused controversy and has already been banned by some countries.

It is obvious that global carbon emissions must be scaled back for economic, social and environmental reasons. Yet it is difficult politically for any one country to take ownership of the issue. China is the world's biggest carbon emitter as well as a world leader in renewables. Fossil fuels have created political instability, so now there is a move towards a new energy structure based on a greater diversity of sources, both geographically and technologically.

19 What is meant by the term energy security?

....

....

20 What is the difference between renewable and recyclable energy?

....

....

....

....

21 State **three** unconventional sources of fossil fuel.

....

22 Describe **one** reason why a government may be reluctant to exploit its unconventional sources of fossil fuel.

....

....

....

23 Describe how a hydrogen fuel cell works.

....

....

....

....

24 Describe **two** reasons why some countries might be reluctant to stop using coal.

....

....

....

....

25 Describe some of the risks associated with energy pathways.

....

....

....

....

26 Explain reasons why renewable sources of energy will not be able to replace fossil fuels in the near future.

....

....

....

....

....

....

....

....

27 Study Table 6.3, which shows the world's ten largest energy consumers in 2019.

Table 6.3

Country	Total energy consumption (Mtoe)	Population (billions)
China	3284	1.439
USA	2213	0.331
India	913	1.380
Russia	779	0.145
Japan	421	0.126
South Korea	298	0.051
Germany	296	0.083
Canada	295	0.037
Brazil	288	0.212
Indonesia	269	0.274

Source: https://yearbook.enerdata.net/total-energy/world-consumption-statistics.html and https://www.worldometers.info/world-population/population-by-country

a   Calculate the range in total energy consumption.

....

....

b   Calculate the percentage difference between the countries with the highest and lowest populations.

....

....

....

c Describe how you would test the strength of the relationship between total consumption and population.

Worked example

A spreadsheet such as Excel could be used to test the relationship between the two variables. The data would be set into two columns and the correlation function used to determine the coefficient, which is a number between −1 and +1. This could also be combined with the use of a null and alternative hypothesis, with critical values allowing the determination of a higher degree of certainty in terms of how likely it is that the relationship is not random.

> Usually students would opt for Spearman's rank as the default tool to test the strength of a relationship. There are other statistical tools available, and a spreadsheet is a rapid and easy way to obtain a correlation coefficient. With Excel you would use the CORREL function.

> You also need to be able to perform Spearman's rank manually, because it can be used as part of the exam on paper 1 or paper 2. It is unlikely that you will have to complete a full test, but instead you might be required to 'fill in the blanks' and to interpret the answer (R_s value).

d Explain **one** reason why the choice of units for population may be inappropriate.

....

....

....

28 Explain **one** reason why natural gas is now used in preference to oil as an energy source.

....

....

....

29 Explain **two** disadvantages of using nuclear power as a source of energy.

....

....

....

30 Explain the advantages and disadvantages of biofuels.

....

....

....

....

31 Explain why OPEC is such an influential player in the world of energy.

....

....

....

....

....

....

32 Explain why a country's energy mix can vary seasonally.

....

....

....

....

....

....

33 Explain the pros and cons of carbon capture and storage.

....

....

....

....

....

....

34 Explain why access to energy is important for countries and their people.

....

....

....

....

....

....

How are the carbon and water cycles linked to the global climate system?

Human threats to biological carbon and water cycles are significant and widespread. Land-use change, linked to deforestation and intensification of agriculture, is an area of geography that you will be familiar with. This, along with ocean acidification and the enhanced greenhouse effect, also threatens these complex and interlinked cycles.

The relationship between people and the carbon and water cycles is intricate, and governments and communities ignore this at their peril. Changes to water and carbon cycles are having adverse impacts on human wellbeing, and resulting in increasing economic costs. Further global warming risks large-scale releases of stored carbon, especially from areas in the Arctic. This will require proactive and extensive responses, such as adaptation strategies and rebalancing the carbon cycle through mitigation. Another urgent need is for global-scale agreements, but achieving these is proving to be problematic due to the 'tragedy of the commons'.

35 What is meant by the term RCP?

....

36 What formula is used to assess the vulnerability of coastal areas to increased environmental threats?

....

....

37 What is meant by the term human wellbeing?

....

....

38 Describe the difference between adaptation and mitigation.

....

....

....

39 Describe **three** impacts that deforestation has on the carbon cycle.

....

....

....

....

40 Describe **two** important effects that global warming is having on the Arctic.

....

....

....

....

41 Describe examples of contrasting attitudes of players to environmental issues.

....

....

....

....

42 Describe **two** land-use changes, other than deforestation, that threaten to change the carbon and water cycles.

....

....

....

43 Describe **three** ways of mitigating global warming.

....

....

....

....

44 Describe **three** impacts that deforestation has on the water cycle.

....

....

....

....

45 Explain how remote sensing can be used to monitor land-use changes.

....

....

....

46 Study Figure 6.2, which shows the effects of deforestation on the carbon cycle.

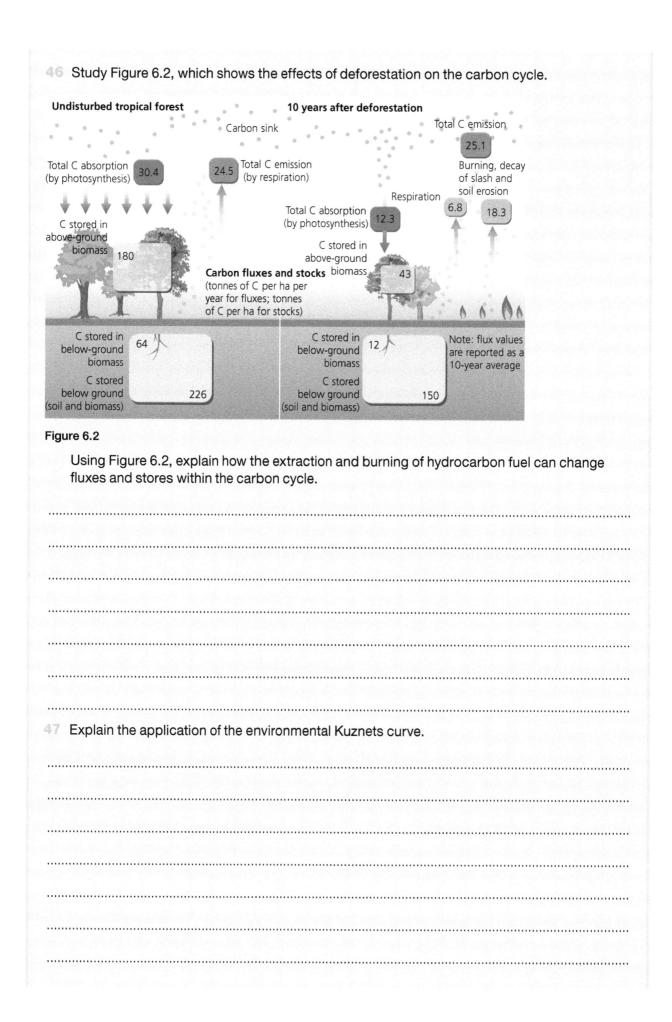

Figure 6.2

Using Figure 6.2, explain how the extraction and burning of hydrocarbon fuel can change fluxes and stores within the carbon cycle.

....

....

....

....

....

....

....

47 Explain the application of the environmental Kuznets curve.

....

....

....

....

....

....

....

....

48 Explain the links between the decline in ocean health and threats to human wellbeing.

....

....

....

....

....

49 Explain the different approaches that can be used to mitigate global warming.

....

....

....

....

....

50 Explain the different attitudes that governments and TNCs have about international agreements relating to carbon emissions.

....

....

....

....

....

51 Explain the main factors creating uncertainty about further global warming.

....

....

....

....

....

....

1 Study Figure 6.3, which shows the carbon cycle for a single tree.

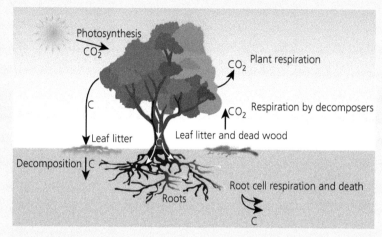

Figure 6.3

a Explain the relationship between photosynthesis in trees and carbon sequestration. (AO1, AO2)

3 marks

④

....

....

....

b Explain the role of oceans in regulating the carbon cycle. (AO1)

6 marks

⑧

....

....

....

....

....

....

c Explain why energy pathways can be easily interrupted. (AO1)

8 marks

⑩

....

....

....

....

....

....

....

....

....

....

Study Figure 6.4, which shows the Kuznets curve.

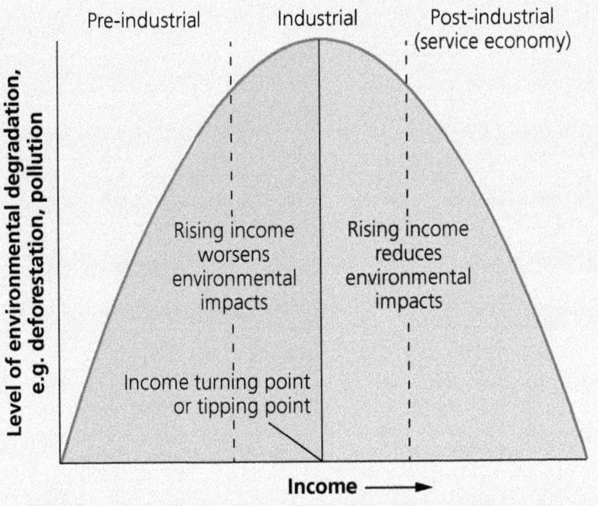

Figure 6.4

d Assess the different factors affecting the position of the tipping point.
(AO1, AO2)

12 marks

15

....

....

....

....

....

....

....

....

....

....

....

....

e   Evaluate the view that, globally, increasing demand for energy is the key
    factor changing the carbon cycle. (AO1, AO2)   **20 marks**

25

....

....

....

....

....

....

....

....

....

....

....

....

....

....

....

....

....

....

....

....

....

....

2   Study Figure 6.5, which shows the effects of deforestation on the carbon cycle.

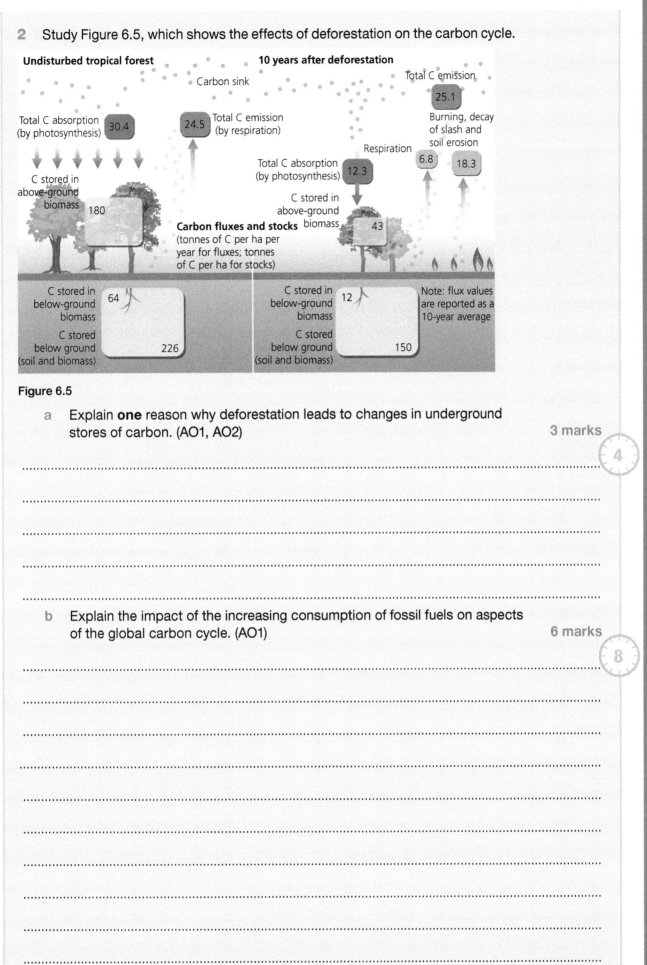

Figure 6.5

    a   Explain **one** reason why deforestation leads to changes in underground
       stores of carbon. (AO1, AO2)

3 marks

....

....

....

....

    b   Explain the impact of the increasing consumption of fossil fuels on aspects
       of the global carbon cycle. (AO1)

6 marks

....

....

....

....

....

....

....

....

....

c Explain how level of economic development influences the energy mix. (AO1)   8 marks

10

....

....

....

....

....

....

....

....

....

....

Study Figure 6.6 which shows four different mitigation strategies.

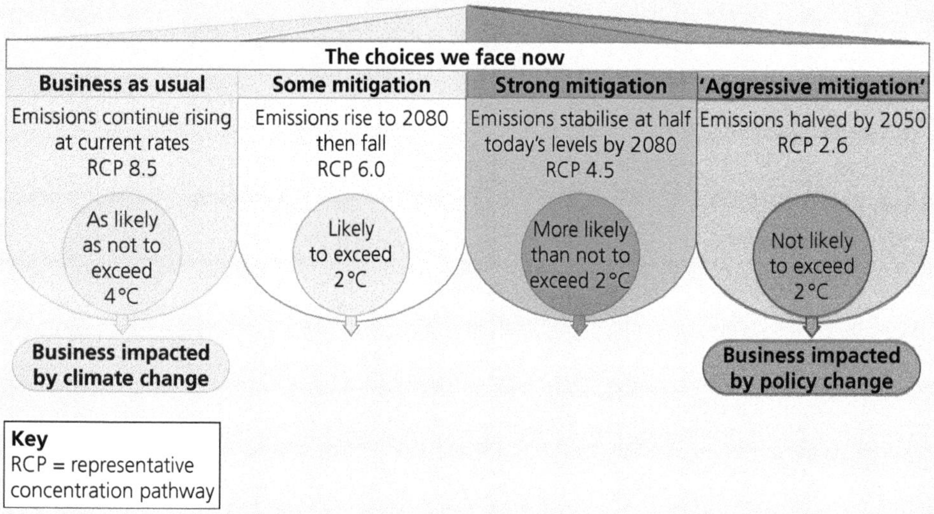

The choices we face now			
Business as usual	**Some mitigation**	**Strong mitigation**	**'Aggressive mitigation'**
Emissions continue rising at current rates RCP 8.5	Emissions rise to 2080 then fall RCP 6.0	Emissions stabilise at half today's levels by 2080 RCP 4.5	Emissions halved by 2050 RCP 2.6
As likely as not to exceed 4°C	Likely to exceed 2°C	More likely than not to exceed 2°C	Not likely to exceed 2°C
Business impacted by climate change			**Business impacted by policy change**

Key
RCP = representative concentration pathway

Figure 6.6

d Assess the role of international governments in determining different mitigation choices. (AO1, AO2)   12 marks

15

....

....

....

....

....

....

....

....

....

....

....

....

....

....

....

....

....

....

....

e   Evaluate the view that renewables are as important as fossil fuels in the global
    energy mix. (AO1, AO2)   20 marks

(25)

....

....

....

....

....

....

....

....

....

....

....

....

Workbook answers at **www.hoddereducation.co.uk/workbookanswers**